T0301832

International Trade in Recyclable and Hazardous Waste in Asia

International Trade in Recyclable and Hazardous Waste in Asia

Edited by

Michikazu Kojima

Institute of Developing Economies, JETRO, Japan

Etsuyo Michida

Institute of Developing Economies, JETRO, Japan

INSTITUTE OF DEVELOPING ECONOMIES (IDE), JETRO

Edward Elgar
Cheltenham, UK • Northampton, MA, USA

Published by
Edward Elgar Publishing Limited
The Lypiatts
15 Lansdown Road
Cheltenham
Glos GL50 2JA
UK

Edward Elgar Publishing, Inc.
William Pratt House
9 Dewey Court
Northampton
Massachusetts 01060
USA

A catalogue record for this book
is available from the British Library

Library of Congress Control Number: 2013934283

This book is available electronically in the ElgarOnline.com
Economics Subject Collection, E-ISBN 978 1 78254 786 0

ISBN 978 1 78254 785 3

Typeset by Servis Filmsetting Ltd, Stockport, Cheshire
Printed and bound in Great Britain by T.J. International Ltd, Padstow

Contents

Contributors

Vella Atienza is an Assistant Professor at the Institute for Governance and Rural Development (IGRD), College of Public Affairs and Development, University of the Philippines (UP), Los Banos, Laguna, Philippines. She previously worked as a Research Fellow at the Institute of Developing Economies, JETRO, Japan from April 2009 to March 2012. Her research focuses on environmental governance in waste management in the Philippines and other Asian countries. She obtained a PhD in Asia Pacific Studies from the Ritsumeikan Asia Pacific University, Beppu, Japan, and a Master of Management and Bachelor of Science in Forestry, both from UP Los Banos. She has authored and edited various publications related to waste management and recycling, forestry, and the environment.

Sungwoo Chung is a Chief Researcher at the Hanwha Life Economic Institute of Korea. He has researched Korean environmental policy including the circulation of recyclable resources. He received a PhD from Hokkaido University, Japan. He worked as a Researcher at the Institute of Developing Economies and Senior Researcher at the Embassy of the Republic of Korea in Japan. He is the co-author of "A study of the WEEE (waste of electrical and electronic equipment)" (in Japanese) (*Journal of the Japan Society of Waste Management Experts* **19** (4), 2008, 235–43).

Michikazu Kojima is a Senior Research Fellow at the Institute of Developing Economies (IDE), Japan External Trade Organization (JETRO). He has studied environmental issues in Indonesia and recycling in the Asian region. He received an MSc from the University of California, Berkeley, in Agricultural and Resource Economics. He is the editor of *International Trade of Recyclable Resources in Asia* (Chiba: Institute of Developing Economies, 2005) and *Promoting 3Rs in Developing Countries: Lessons from the Japanese Experience* (Chiba, Institute of Developing Economies, 2008).

Etsuyo Michida is an Associate Senior Research Fellow at the Institute of Developing Economies, JETRO. She earned a PhD in Economics from Kobe University. Her work area is the environment and trade, including recyclables trade and cross-border chemicals management. Her major work is "North–South trade and industry-specific pollutants," *Journal of Environmental Economics and Management*, **54**, 229–43.

Shozo Sakata is a Senior Research Fellow at the Institute of Developing Economies, JETRO, Japan. He has worked on broad issues relating to socio-economic development in Vietnam. His research topics range from rural development, environmental, and labour issues to the development of local enterprises in Vietnam. He is the editor of *Actors for Poverty Reduction in Vietnam* (Institute of Developing Economies, 2006) and *Vietnam's Economic Entities in Transition* (Palgrave Macmillan, 2013).

So Sasaki is an Associate Professor at the Faculty of Economics, Chuo University. He has studied waste management and recycling policy in Thailand and environmental business in the Asian region. He received a PhD from Hokkaido University in Environmental Economics. He is the co-author of "Difficulties in applying extended producer responsibility policies in developing countries: case studies in e-waste recycling in China and Thailand," *Journal of Material Cycles and Waste Management*, **11** (3), 2009, 263–9.

Tadayoshi Terao is a Senior Research Fellow at the Institute of Developing Economies, JETRO. He received his Master's degree in Agricultural Economics from the University of Tokyo. He has studied industrial pollution, environmental policies, social movement, and agricultural policies in East and South Asia. He authored "An institutional analysis of environmental pollution disputes in Taiwan, cases of 'self-relief'" (*The Developing Economies*, **XL** (3), 2002), and co-edited the book *Development of Environmental Policy in Japan and Asian Countries* (Palgrave Macmillan, 2007).

Jun Tsuruta is an Associate Professor of International Law and Maritime Environmental Law at the Japan Coast Guard Academy, Hiroshima, Japan. He earned a Master's degree in Law (International Law) from the Graduate School of Law and Politics, the University of Tokyo. He has published many articles on the law of treaties, international law of the sea, and international environmental law, in Japanese and English. His recent works in English include "Japanese measures against marine pollution under UNCLOS and the IMO treaties," *Journal of East Asia and International Law*, **2** (2), 2009, and "The Japanese Act on the Punishment of and Measures against Piracy," *The Aegean Review of the Law of the Sea and Maritime Law*, **1** (2), 2011.

Aya Yoshida is a Senior Researcher at the National Institute for Environmental Studies. She received a PhD from the Department of

Urban Engineering at the University of Tokyo in 2006. She is author of "China: the world's largest recyclable waste importer," Chapter 3 in *International Trade of Recyclable Resources in Asia* (Chiba: Institute of Developing Economies, 2005).

Preface

The international trade of recyclable waste has expanded in Asia in recent years. Western countries also have exported huge amounts of waste paper, waste plastics, metal scraps, and other material to Asia. Recyclable waste has helped meet the growing demand for resources that has accompanied rapid economic growth in the Asian region. Economic integration of the region also has led to an increase in the volume of international trade of recyclable waste.

Improving resource efficiency through recycling is emphasized in various international initiatives including the International Resource Panel organized by United Nations Environment Programme and the G8 3R Initiative. While international recycling with environmentally sound technology may be able to contribute resource efficiency, importing countries such as China have been negatively affected by pollution from the recycling industries utilizing the imported waste. On the other hand, the recycling industries with environmentally sound technology in exporting countries face a lack of recyclable waste. Various countries have started to regulate the international trade of recyclable waste in order to maximize the economic benefit of and to minimize negative impact from international recycling. Unilateral efforts tackling these issues on international recycling have not been very effective thus far because of weak enforcement of border controls, limited cooperation between importing and exporting countries, and a lack of knowledge of the characteristics of recyclable waste and the recycling process. These issues should be solved internationally, based on a common understanding of international recycling in Asia.

I have had many chances to talk about the issues related to international recycling with experts, government officers, businessmen, and staff in nongovernmental organizations in Asian countries. I always feel that these stakeholders do not know much of the basic information about international recycling, such as the volume of international trade of recyclable waste, the problems caused by international recycling, and the struggles to properly manage the trade. As a result, there is a lack of common understandings and only limited discussion about creating a society in Asia with a sound material cycle and recycling system. I hope this book facilitates a common understanding of the issues of international recycling among Asian countries, which can spur effective international and regional cooperation needed to establish a sound material cycle society not just in Asia but worldwide.

In the process of making this book, the authors conducted many interviews with government officers, experts, and businessmen in various countries. On behalf of all of the project members, I would like to thank the many persons who have given us kind support and advice in this research project.

Michikazu Kojima
Senior Research Fellow
Institute of Developing Economies, JETRO

1. Issues relating to the international trade of second-hand goods, recyclable waste, and hazardous waste

Michikazu Kojima

INTRODUCTION

Economic integration has progressed considerably in Asia, and most Asian countries have seen an increase in trade dependency, which is calculated as the total value of exports and imports divided by GDP. Economic integration also stimulates economic growth in the region. With deepening economic integration and accelerating economic growth, the international trade of recyclable waste has also increased to fulfill the growing demand for resources. Hazardous wastes are also traded for disposal and recycling. However, the increase in international trade of recyclable waste and hazardous waste consequently generates environmental problems such as unsound disposal practices and pollution from recycling processes.

To prevent these problems, various trade regulations have been enacted. In particular, the Basel Convention regulates the transboundary movement of hazardous waste, which is defined in the annex of the convention. Additionally, some countries also maintain their own list of hazardous waste types, and have introduced regulations on the import of recyclable waste and second-hand goods. Such regulations include standards governing recyclable waste import and pre-shipment inspections, and import bans on second-hand goods.

This chapter reviews the issues relating to the international trade of second-hand goods, recyclable waste, and hazardous waste and gives an overview of the book as a whole. Section 1.1 reviews the issues relating to the international trade of recyclable waste and hazardous waste historically. Section 1.2 overviews trade measures implemented to prevent and mitigate the issues presented in the previous section. Section 1.3

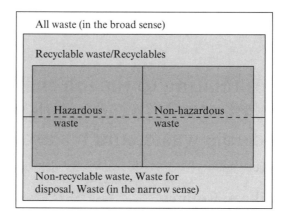

Source: Created by the author.

Figure 1.1   Definition of waste

summarizes the current issues relating to the international trade of recyclable waste and hazardous waste, and Section 1.4 outlines the structure of this book.

Before looking at the main argument, we will introduce some terminology used throughout the book. In our usage, recyclable waste and recyclables have the same meaning, even though in some languages, such as Vietnamese, "waste" does not include "recyclable waste" and refers only to "waste for disposal." "Hazardous waste" is waste that has hazardous characteristics and includes explosives, flammable liquids or solids, corrosives, and toxic substances which are defined in Annex III of the Basel Convention on the Control of Transboundary Movements of Hazardous Wastes and their Disposal. A portion of hazardous waste is recyclable, the rest is non-recyclable. Figure 1.1 illustrates the relationship between these various terms. Lastly, we use the terms "transboundary movement" and "international trade" interchangeably.

Concerning the data used in several chapters of this book, we use two main types of trade statistics. The first type is trade statistics compiled by customs officials and other similar authorities. The codes for classifying these data are for the most part harmonized by the World Customs Organization through the Harmonized Commodity Description and Coding System (HS code). The other type of trade statistics is the data on hazardous waste trade with prior notice and consent, which is compiled by parties and reported to and published by the Secretariat of Basel Convention or disclosed to the public by some governments.

1.1   A HISTORICAL OVERVIEW OF THE INTERNATIONAL TRADE OF RECYCLABLE WASTE AND HAZARDOUS WASTE

Recyclable waste is not a commodity new to international trade. The international trade of recyclable waste is recorded in Japanese trade statistics from the 19th and early 20th centuries: for example, 68 tons and 3,482 tons of copper scrap were exported in 1868 and 1871, respectively, and in 1917, 11 thousand tons of waste cotton were exported, with a value of 3.17 million yen.[1] This price of cotton waste was more than double that of rice at the time. Cotton waste was used to make products for wiping up oil. At that time, recyclable waste and scraps were internationally traded in the same manner as other goods.

Until the 1970s, studies ignored the negative externalities of pollution from international recycling. However Grace et al. (1978), using a partial equilibrium model with the assumption that demand and supply of recyclable wastes differed between two countries, were the first to show benefits to the environment in economic terms. In the 1980s, a number of incidents occurred where hazardous waste was exported from developed countries to developing counties, and the environment and human health were harmed in the importing country as a consequence. For example, hazardous waste was exported from Italy to Koko, Nigeria, where the hazardous chemicals contaminated the surrounding environment. These incidents were comprehensively reviewed by the Center for Investigative Reporting and Bill Moyers (1990). In the 1990s, Copeland (1991) developed a theoretical model on international trade of waste for disposal. It shows that if the market is perfect and if the disposal industry is regarded the same as other manufacturing industries, there is no justification for hazardous waste control and trade restriction. However, assuming that the government will not be able to implement proper regulation of the waste disposal industry and assuming the non-existence of optimal taxes on illegal dumping, trade restriction can be the second best policy.

The Basel Convention on the Control of Transboundary Movements of Hazardous Wastes and Their Disposal (hereafter, the Basel Convention) was drafted to prevent improper transboundary movement of hazardous waste that can harm the environment and human health in importing countries. The convention entered into effect in 1992. Prior notice and consent are required before the shipment of hazardous waste. In the late 1980s and early 1990s, the number of hazardous waste treatment and disposal facilities was limited. At the Second Conference of Parties (COP) in 1994, a ban on the export of hazardous waste from developed countries to developing countries for disposal was adopted. However, the transboundary movement of waste for the purpose of recycling was permitted, and the decision of the COP was only to encourage developing countries to ban the import of hazardous waste for recycling. At the third COP in 1995, an amendment was adopted that banned developed countries (annex VII countries: EU countries, OECD countries, and Liechtenstein) from exporting hazardous waste to developing countries. The decade after the Basel Convention went into effect was a period in which trade regulations were made stricter. Various studies from 1990 to 2004 focused on the legal aspects of the Basel Convention and the pros and cons of the Ban amendment (Kumar 1996; Asante-Duah and Nagy 1998; Basel Action Network Silicon Valley Toxic Coalition 2002).

Around 2005, this trend started to change. A substantial amount of non-hazardous recyclable waste was shipped from the European Union, Japan, and United States to China, India, and other countries for recycling. Hazardous waste recycling facilities with environmentally sound technologies were established in middle- and high-income countries such as Singapore, Malaysia, Thailand, and other non-annex VII countries. The G8 adopted the 3R initiative at the Sea Island Summit in 2004, which described the goal of reducing barriers to the international trade of recyclable waste and recycled goods as follows: "Reduce barriers to the international flow of goods and materials for recycling and remanufacturing, recycled and remanufactured products, and cleaner, more efficient technologies, consistent with existing environmental and trade obligations and frameworks." Some countries started easing import restrictions of recyclable waste. China, which had prohibited the import of unwashed and uncrushed waste polyethylene terephthalate (PET) bottles, began to allow the import of such bottles under certain conditions. Indonesia had prohibited the import of plastic waste after dozens of shipping containers holding such waste had been left in Indonesian ports; however, in 2008 Indonesia began to allow its import following expansion of the plastic recycling industry.

Economic integration, fragmentation of production processes, and the availability of technology have all been important factors contributing to the need to ease import restrictions on recyclable waste. With deepening economic integration in Asia, numerous products, parts, and materials are now made in various countries and internationally traded. The international trade of recyclable waste is not an exception to this trend (Kojima 2005). Recycling industries for specific waste types are not located in every country, and thus trade is necessary. For example, Singapore does not have recycling facilities for lead acid batteries or paper mills that utilize waste paper. Accordingly, these materials are exported to other countries with such capabilities.

Some countries and NGOs are, however, still anxious about improper dumping of hazardous waste and the receipt of other waste from developed countries by developing countries. In the Ministerial Conference of the 3R Initiative held in Tokyo in 2005, ministers discussed the future actions of the G8 3R initiative and some developing countries expressed concern over the easing of trade barriers on hazardous waste and recyclable waste. In addition, when the Japanese government tried to make Economic Partnership Agreements (EPAs) with countries such as the Philippines and Thailand, various non-governmental organizations (NGOs) opposed the EPAs since the agreements were perceived as promoting the free trade of hazardous waste.[2] Some HS codes under which hazardous waste is classified are listed as goods for zero tariff. Although the EPAs also state that both parties should follow other international treaties, including the Basel Convention, some NGOs still consider the EPAs to be inappropriate.

At the tenth meeting of the COP, held in Cartagena, Columbia in October 2010, the omnibus decision of country-led initiatives was adopted. This decision affects the future directions of the Basel Convention. At the COP, it was decided that the legal interpretation of paragraph 5 of Article 17 required amendment, and the conditions for the amendment entering into effect were subsequently made clear.[3] Other important decisions covered in the omnibus decision are "Developing Standards and Guidelines for ESM" (environmentally sound management) and "Providing Further Legal Clarity." The Basel Convention developed various guidelines for ESM and legal interpretations of terminology. However, a more systematic approach was required and comprehensive efforts were made to improve the requirements for ESM. Moreover, the unclear distinction between waste and non-waste for certain used equipment and second-hand goods hindered enforcement of the regulation for suspicious shipments. The activities provided for by "Developing Standards and Guidelines for ESM" and "Providing Further Legal Clarity" are expected to enable the convention to be effectively enforced.

1.2 TRADE MEASURES ON RECYCLABLE WASTE, HAZARDOUS WASTE, AND SECOND-HAND GOODS

The international trade of recyclable waste and second-hand goods may conserve resources and be economically beneficial. However, in connection with the international trade of those goods, some problems have emerged. One problem is inadequate recycling processes generating pollutants. The recycling of lead in Taiwan in the late 1980s is a representative case of this problem. Waste lead acid batteries were imported to Taiwan from the United States, Japan, and other countries, and inadequate pollution controls led to lead pollution that harmed children living near the recycling facilities. The affected children exhibited lower intelligence levels upon entering kindergarten. Another problem is that imported waste is sometimes disposed of without any treatment at the destination. For example, nearly 3,000 tons of hazardous waste containing mercury were shipped from Taiwan to Sihanoukville, Cambodia, and dumped improperly. It is reported that workers and residents experienced negative health effects including diarrhea, headache, fever and skin rashes.[4] Second-hand goods can also create problems. For instance, second-hand vehicles that do not meet emissions standards increase air pollution and older second-hand vehicles in general are likely to emit more pollutants than new ones. Furthermore, the life span of imported second-hand goods may be shorter than that of new products, and thus they become waste sooner. For example, the use of second-hand imported electronics may increase the volume of electronic waste in the importing country. To prevent these types of problems, some countries have completely banned the import of second-hand goods, while others allow imports of second-hand goods that were manufactured within a certain number of years. Examples of trade measures implemented by Asian countries are presented in Table 1.1.

As listed in the annex of the Basel Convention, the regulated hazardous waste types are defined according to several aspects. The first aspect is the "waste stream" such as "clinical waste from medical care in hospitals, medical centers and clinics," "waste from the production and preparation of pharmaceutical products," and "waste from the production, formulation and use of organic solvents." The second aspect is "waste having as constituents" including "metal carbonyls," "arsenic; arsenic compounds," and "cadmium; cadmium compounds," among others. The third aspect is hazardous characteristics such as whether the material is "explosive," "flammable," "oxidizing," "toxic," or "ecotoxic." Regulated hazardous waste types are also listed in Annex VIII of the convention. However, testing methods and minimum concentration levels are not defined in the convention and are left to each country to define. As a result, each country has its own definition of hazardous waste, and the differences in these definitions can cause misunderstanding between exporting and importing countries and lead to the unintentional illegal trade of hazardous waste.

Some countries also prohibit the import of specific recyclable waste types that are not listed in the Basel Convention. For example, Indonesia prohibited the import of waste plastic, because it was imported and abandoned in several Indonesian ports in the early 1990s. Also, Thailand banned the import of waste tires in 2003 after no-one used the imported

Table 1.1   Trade measures governing hazardous waste, recyclable waste, and second-hand goods

Trade measure	Content	Example (related chapters)
Prior notice and consent	The exporter and/or importer should make prior notice and consent before shipment	The Basel Convention requires the parties to follow prior notice and consent mechanism (Chapters 6, 7, 8, 9)
Standards for importing recyclable waste	Recyclable waste to be imported should comply with quality standard set by the government	China has various standards for importing recyclable waste such as plastics and mixed metals (Chapters 3, 4 and 5)
Ban on import of specific type of waste	Import of specific type of waste is prohibited	China, Vietnam and Indonesia prohibit importing hazardous waste (Chapters 3 and 4)
Ban on export of hazardous waste to developing countries	Export of hazardous waste is prohibited to developed countries	The Ban Amendment of the Basel Convention requires developed countries to prohibit exporting hazardous waste to developing countries
Pre-shipment inspection	Importing country requests exporter to undergo pre-shipment inspection. Inspector checks the quality of waste	The Chinese government mandates pre-shipment inspection of second-hand goods and recyclable waste such as used paper, waste plastic, and steel scraps. Foreign exporters of used paper to Indonesia must undergo pre-shipment inspection, to prevent the import of banned printed materials that are viewed as religiously problematic (Chapter 3)
Tariff on second-hand goods according to year of production	Special tariff or tax on older second-hand equipment is imposed	Mongolia imposes a special tax on imported second-hand automobiles, depending on the age of the automobile

Table 1.1 (continued)

Trade measure	Content	Example (related chapters)
Registration of importers	Importer must register. If importer violates the law, the license will not be renewed	China introduced a registration system for importers of recyclable waste. Singapore also has a registration system for importers of recycled tires (Chapter 3)
Registration of exporters	Exporter in foreign country must register. If exporter violates the law, the license will not be renewed	China introduced a registration system for exporters of recyclable waste to foreign countries in January 2005 (Chapter 3)
International manifests	Manifest system where the importer must send back a manifest to the exporting county	Taiwan set up an Internet-based manifest system, in which recyclers in foreign countries must provide documentation when they receive recyclable waste and when they treat the material
Inspection of treatment facilities in foreign countries	Inspection of treatment facilities in foreign countries is conducted	Taiwan sends specialists to treatment facilities in foreign countries that import and treat hazardous waste from Taiwan. The inspectors also conduct interviews of local and central government officials in the importing country
Reporting requirement for importers and/or exporters	Importer should send report to the government	Export/Import Declaration System in South Korea requires importer and exporter to report on the import and export of non-hazardous waste (Chapter 5)

Source: Compiled from various sources.

waste tires that were abandoned in port areas. China has established standards for imported recyclable waste. In regard to waste PET bottles, crushed and washed PET flakes are allowed to be imported, whereas baled waste PET bottles were prohibited until only recently. The justification of this regulation is the possibility of contamination by residues inside the bottles. Pre-shipment inspection is also required to check whether recyclable waste to be imported satisfies the standard. Exporters of recyclable waste to China arc required to undergo pre-shipment inspection in the exporting country. The Indonesian government also requires exporters of used paper to undergo pre-shipment inspection in order to prevent the import of religiously problematic documents such as pornography.

Tariffs can be used to control the quality of imported recyclable waste and second-hand goods. For example, Mongolia imposes a special tax on imported second-hand automobiles, with higher taxes being levied on older automobiles. Registration systems for importers and exporters have also been established, and importers must have facilities that meet certain criteria. China introduced a registration system not only for importers but also exporters intending to export recyclable waste to China. If an exporter does not have a suitable environmental management system, or violates the regulations, the permit is not renewed.

To ensure proper treatment of exported hazardous waste, Taiwan introduced a manifest system or consignment note system to cover both domestic and international transactions. Importers of hazardous waste sent from Taiwan should input the information in the Internet-based manifest system when receiving and treating the materials. Taiwan also sends inspection teams to foreign countries.

1.3   CURRENT ISSUES RELATING TO THE INTERNATIONAL TRADE OF HAZARDOUS WASTE AND RECYCLABLE WASTE

If the externalities resulting from recycling and waste disposal are internalized, in other words, if pollution control regulations on the recycling industry and disposal services are enforced effectively, the trade measures described in the previous section are not necessary. However, in most developing countries, pollution control regulations are not fully enforced and the technical capabilities to control pollution are still limited. Governance is still weak in many developing countries, and the trade measures that have been introduced take into account the weak enforcement and governance in these countries. Developing countries also need resources for economic development. Recyclable waste can contribute to meeting the increasing demand for resources. Further, developing countries have comparative advantages in labor intensive processes, such as dismantling and sorting. Thus, developing countries use trade regulations to minimize environmental pollution from the import of recyclable and hazardous waste and maximize the use of recyclable and hazardous waste as resources.

On the other hand, developed countries such as Japan have made efforts to develop a society with a sound material use cycle. Legislation to promote recycling has been established in many developed countries and numerous companies have invested in the recycling business. However, some recycling industries face shortages of recyclable and hazardous waste, because the export of these wastes and used items containing valuable material is increasingly common (Working Group on Enhancing International Recycling, Industrial Structure Council/METI 2004). In addition, because of increasing resource prices and instability in the supply of rare metals, concerns over resource efficiency have increased.

Some developing countries also have tried to prevent waste problems such as improper disposal and recycling. The Philippines enacted the Ecological Solid Waste Management Act in 2001, which attempts to reduce waste through recycling. The act also mandates government agencies to facilitate the recycling industry. However, a significant amount of recyclable waste is exported to China and other Asian countries, some of which may be sent to improper recycling facilities.

The major challenge then is to both prevent the illegal and illicit transboundary movement of hazardous waste and promote the use of recyclable waste internationally. Trade regulations are needed to prevent such illegal trade, although overly tight trade regulations often hinder legitimate international trade where waste is destined for recycling facilities with environmentally sound technologies.

As resource scarcity has become an important issue, the International Resource Panel was established in 2007 by the UNEP to develop holistic approaches to the management of global resources. The efficient use of resources including recyclable waste has been among the topics discussed. Therefore, the challenge that we face is how to devise appropriate trade measures, prevent improper dumping and recycling, and facilitate proper transboundary movement to improve resource efficiency on a national, regional, and global scale. Although some countries with few or no facilities for the treatment of recyclable waste and hazardous waste may be reluctant to ease trade regulations, moving forward, it is important to review and analyze the existing regulations and the background of illegal and illicit transboundary movement of hazardous waste, to find appropriate trade measures and other policies.

1.4   STRUCTURE OF THIS BOOK

This book consists of three parts. Chapters 1 and 2 provide an overview of recycling issues in the era of economic integration and the trade flows of recyclable waste in Asia, respectively. Chapters 3 through 8 discuss trade regulations on recyclable waste and the impact of international trade on recycling in selected Asian countries. These countries are struggling to manage the negative impact on the environment of recycling traded scrap and to maximize resource utilization. Chapter 3 focuses on the situation in China, the biggest importer of recyclable waste. The Chinese government is trying to prevent hazardous waste import, to control the quality of imported recyclable waste, and to utilize imported recyclable waste as resources. China is aggressively applying various trade measures to minimize the problems from imported recyclable waste and maximize resource utilization. Chapter 4 focuses on Vietnam, which has increased its imports of recyclable waste significantly in the last ten years. The author explores the "dilemma" from the perspective of the increasing import of recyclables, which provides valuable resources for development but at the same time causes pollution problems. Chapter 5 highlights the situation in South Korea, which is trying to manage imported recyclable waste, such as coal ash, through the use of a newly established reporting system. Also, since South Korea is an exporter of some recyclable waste to China, it must also try to control illegal exports of certain hazardous waste. In addition to the implementation of the Basel Convention, South Korea has introduced the Export/Import Declaration System in 2008 in an attempt to grasp the international trade of non-hazardous waste. Chapter 6 highlights the impact of economic integration on the domestic recycling system in the Philippines. As mentioned previously, the Philippines enacted legislation in 2001 on the management of solid waste, which requires the government not only to make the waste management system environmentally sound, but also to facilitate recycling and waste reduction. However, because of the weak industrial infrastructure and the high demand for recyclable waste in the region, exports of recyclables have increased, which undermines the development of the recycling industry in the Philippines. Chapter 7 discusses the national law implementing the Basel Convention in Japan. Specifically, we look at the difficulty of detecting hazardous waste concealed by using forged documents. To address this issue, the author proposes stricter punishments, such as establishing criminal penalties for the "attempt and preparation" of exporting hazardous waste. Chapter 8 discusses the impact of import duty reduction systems on the international trade of hazardous waste and recyclable waste. Many developing Asian economies enact measures to promote export-oriented industrialization. Such measures include the establishment of export-oriented industrial zones and reductions in import duty for manufacturing products to be exported. It is easier to export production waste generated in export-oriented zones than to send it to a domestic recycling company. Although some waste cannot be treated in the country in which it was generated, domestically recyclable waste is also exported to other countries, which can be seen as an economic loss.

Chapters 9 and 10 of the book focus on the international management system. Chapter 9 reviews the issue of illegal traffic in the Basel Convention through case studies of illegal shipments in Asian countries. This chapter points out the need for further information exchange in the definitions of hazardous waste and joint efforts to detect the flow of illegal shipment among Asian countries. Chapter 10 examines the subject of ship recycling. In the past 60 years, the center of ship dismantling has shifted from Japan and Taiwan to South Asian countries and China, and this chapter discusses the reasons for this shift. Chapter 11 recaps the previous chapters, offers our conclusions, and discusses future actions for the international community.

NOTES

1. Toyo Keizi Shinpo Sha (1935).
2. An example of their views is the Basel Action Network (2007).
3. A fixed time approach was adopted. The decision at COP 10 stated that "the acceptance of three-fourths of those parties at the time of the adoption for the amendment is required for the coming into force of such amendment."
4. Human Rights Watch (1999).

REFERENCES

Asante-Duah, De Kofi and Imre V. Nagy (1998), *International Trade in Hazardous Waste*, London: E & FN SPON.
Basel Action Network (2007), *JPEPA as a Step in Japan's Greater Plan to Liberalize Hazardous Waste Trade in Asia*.
Basel Action Network and Silicon Valley Toxic Coalition (2002), 'Exporting harm: the high-tech trashing in Asia'.
Center for Investigative Reporting and Bill Moyers (1990), *Global Dumping Ground*, Center for Investigative Reporting.
Copeland, Brian R. (1991), 'International trade of waste products in the presence of illegal disposal', *Journal of Environmental Economics and Management*, **20**, 143–62.
Grace, Richard, R. Kerry Turner and Ingo Walter (1978), 'Secondary materials

and international trade', *Journal of Environmental Economics and Management*, **5**, 172–86.

Human Rights Watch (1999), *Toxic Justice: Human Rights, Justice, and Toxic Waste in Cambodia*, downloaded from http://www.unhcr.org/refworld/docid/3ae6a80012.html (accessed 16 April 2012).

Kojima, Michikazu (ed.) (2005), *International Trade of Recyclable Resources in Asia*, Institute of Developing Economies, http://www.ide.go.jp/English/Publish/Spot/29.html.

Kumar, Katharina (1996), *International Management of Hazardous Wastes: The Basel Convention and Related Legal Rules*, London: Oxford University Press.

Toyo Keizai Shinpo Sha (1935), *Foreign Trade of Japan: a Statistical Survey*.

Working Group on Enhancing International Recycling, Industrial Structure Council METI (2004), *Toward a Sustainable Asia Based on the 3Rs*, http://www.meti.go.jp/policy/recycle/main/english/council/reports/report_sutainableasia_en.pdf.

2. International trade of recyclables and policies to support their sustainable use in Asia

Etsuyo Michida

INTRODUCTION

The international trade of recyclables has surged in the Asian region. Recyclable resources such as metal scraps, waste paper, and waste plastics, which were underutilized in the countries of generation, are now being placed into the goods production cycle, being processed and recycled, as important resources in other countries. Given that trade in recyclable resources has increased in recent decades, the cross-border utilization of these resources has become an important issue and is expected to become even more important in the future considering the constraints on the supply of natural resources. It is in our interests then to examine the situation and characteristics of the trade of recyclables in Asia and to learn the drivers of the trade.

The increase in the trade flow of recyclables and in the utilization of the recyclable resources across borders has not been achieved without side effects: imported recyclables and the illegal import of hazardous waste to developing countries have caused environmental problems in some cases. Countries have amended or changed their trade and environmental regulations[1] in efforts to combat such problems and to control environmental pollution caused by the trade of recyclables. In recent decades, however, it has become increasingly clear that these trade and environmental policy changes can also hamper the smooth trade flow of recyclables. Acknowledging that it is inevitable that we better utilize recyclable resources across borders, we need to consider how policies can maintain a balance between improving resource efficiency and protecting the environment. To consider this issue, we must examine the role of the trade and environmental policies enacted in different countries and their impacts. The interaction between policy and waste trade has previously been addressed in the literature. Copeland

(1991) considers the restriction of waste trade when illegal trade can be conducted with a theoretical model. Levinson (1999) empirically examines the impact of taxation on hazardous waste trade, and Ley et al. (2000) focus on the interaction between municipal waste trade and trade restriction policy in the United States. Issues related to the trade of recyclables involve different aspects from those of non-useable waste because recyclables are processed and used as materials for products in importing countries. With regard to the trade of recyclables, Ray (2008) discusses issues in Asia and points out that tightened regulations could create vulnerability in the trade of recyclables. Fujii (2010) has focused on the recyclable trade in China and shows that China has increased its import of recyclables and become a leading importer of recyclables in the world. It describes the structure of importation of recyclables in China and East Asia as a factor driving exports for the country. This chapter also focuses on recyclables and elaborates further on the interaction between trade and policies regulating environmental pollution based on some cases observed.

Section 2.1 describes trends in the trade flow of recyclables in Asia.[2] Section 2.2 discusses the factors that affect such flow. In Section 2.3, after examining the characteristics of importation for some recyclables on a geographic basis, the factors that determine the location of these industries are discussed. Section 2.4 shows how Asian governments implement trade and environmental regulations related to imported recyclables. Then the impact of these regulations on trade flow is examined. Section 2.5 discusses the policy implications.

2.1 THE TRADE OF RECYCLABLES IN ASIA

A rise in demand for resources has driven an increase in the cross-border trade of recyclables such as iron, copper, and aluminum scrap, as well as waste paper and plastic.[3] From 1992 to 2010, in Asia, waste plastic trade increased more than thirty-fold and trade in metal scrap increased by about six-fold for copper scrap and three-fold for iron scrap in the same period (Figure 2.1). Backed by a large demand for raw materials, Asia is now a center for recyclable imports; in 2010, Asia took around 80 percent of the world's import of copper scrap, 80 percent of waste plastic, 60 percent of waste paper, and 30 percent of iron and steel scrap.[4]

The importance of traded recyclables cannot be underestimated: the volume of such trade has become a significant portion of the overall resource trade, which includes the trade of virgin resources, recyclable

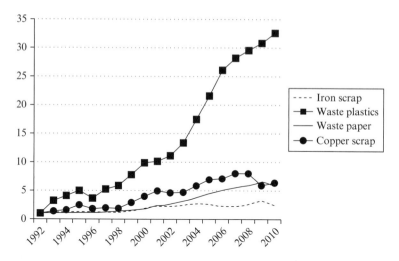

Note: Asia refers to China, Hong Kong, India, Indonesia, Japan, Malaysia, Philippines, Singapore, South Korea, Thailand, Taiwan, and Vietnam.

Source: Created by the author using UN Comtrade data, Taiwan Trade Statistics, and World Trade Atlas.

Figure 2.1   Increase in recyclables trade (quantity) in Asia from 1992 to 2010 (1992=1)

waste to be recycled, and recycled resources that are processed and ready for use in production. In China in 2010, iron and steel scrap made up 22 percent of imports in terms of quantity for the iron and steel resource trade (HS72) and accounted for 49 percent of the copper and articles imported (HS74). Waste paper accounted for 70 percent of the import of raw materials for paper including wood pulp and cellulosic fiber (HS47).[5]

The importation of recyclables has increased in emerging countries in Asia in recent decades, due to the high demand from these countries for such goods. Table 2.1 shows the imports of recyclables in 2010 for individual countries. While China has the largest demand for waste paper, waste plastic, and copper scrap in the region, the ASEAN 6 countries import a large volume of iron and steel scrap. South Korea and Taiwan are also major importers of iron and steel scrap.

Are recyclable resources traded mainly within the Asian region? Table 2.2 shows Asia's shares of recyclable imports as well as exports of different resources for each country. For example, in regard to the waste paper trade of Japan, 97 percent of its waste paper export is destined for

Table 2.1   *Imports of recyclables in Asian countries in 2010 (unit: 1,000 tons)*

	Waste paper	Waste plastic	Iron and steel scrap	Glass cullet	Copper scrap	Precious metal scrap	Aluminum scrap
Japan	44	2	491	12	159	44	77
S. Korea	1,122	47	8,089	79	203	10	546
Taiwan	569	149	5,366	7	90	NA	163
China	*24,538*	*12,805*	*6,200*	*2*	*4,519*	*.*	*2,906*
Mainland	24,352	8,010	5,848	2	4,364	.	2,854
Hong Kong	186	4,795	352	.	155	.	52
ASEAN6	*3,814*	*199*	*5,526*	*141*	*43*	*3*	*202*
Indonesia	2,412	40	1,642	52	14	.	38
Philippines	120	1	29	4	2	.	1
Malaysia	180	82	2,292	129	13	1	64
Singapore	78	4	258	.	4	2	4
Vietnam	.	54	23	NA	.	NA	.
Thailand	1,024	18	1,282	8	10	.	95
India	*2,544*	*84*	*5,137*	*127*	*108*	*NA*	*524*

Note: Period (.) in the table refers to values less than 500 tons.

Source: UN Comtrade and World Trade Statistics.

other Asian countries, while only 3 percent of waste paper is imported from Asia. The statistics show that countries in Asia import not only from Asia but also from outside the region. This suggests that procurement of recyclables in Asia depends on countries both in and outside the region. For certain types of resources, Asia has served as the world center for recycling.

The significance of such imports and their growth can be highlighted in the waste plastic and waste paper imports of China (Figures 2.2 and 2.3). These imports show that China demands recyclable resources from all over the world. The majority of the exported waste plastic from the European Union, the United States, and Japan went to China and Hong Kong. In terms of waste paper, more than half of that exported from the European Union, the United States, and Japan is shipped to China and Hong Kong. Moreover, the exported amount of these two resources increased significantly between 1997 and 2007. Clearly the economic areas of the European Union, the United States, and Japan increased their dependence on China for recycling these recyclables.

Table 2.2   Asia's share of import and export of recyclables in 2010 (%)

		Waste plastic	Waste paper	Iron and steel scrap	Copper scrap	Aluminum scrap
China	Export	31	82	32	72	25
	Import	55	20	56	18	100
Japan	Export	95	97	95	100	99
	Import	91	3	46	63	26
South Korea	Export	100	83	68	98	89
	Import	84	10	43	28	15
Taiwan	Export	98	99	84	100	100
	Import	70	30	22	58	3
Indonesia	Export	99	92	81	90	100
	Import	10	26	23	68	33
Malaysia	Export	89	100	17	86	86
	Import	26	25	13	41	61
Philippines	Export	89	99	82	92	98
	Import	97	36	69	54	38
Singapore	Export	92	89	78	92	73
	Import	72	81	36	62	71
Thailand	Export	95	91	77	94	62
	Import	48	57	9	53	16
Vietnam	Export	94	86	N/A	N/A	100
	Import	56	46			8

Notes:
(1)   Asia refers to the ASEAN6 (Indonesia, Singapore, Thailand, Malaysia, Philippines, and Vietnam), Japan, China, South Korea, and Taiwan.
(2)   N/A refers to data not available.
(3)   China refers to mainland China.

Source: UN Comtrade and World Trade Atlas for Taiwan.

2.2   FACTORS AFFECTING THE TRADE FLOW OF RECYCLABLES

Recyclables without sufficient demand in the countries of generation are shipped to other countries and are utilized more efficiently at major destinations in less developed countries. There are several factors behind the surge in the trade of recyclables in Asia.

The first factor relates to the recovery processes for recyclables. Current practices and technology make this work labor intensive. When collected, many recyclable resources such as metal scraps, waste plastic, or waste

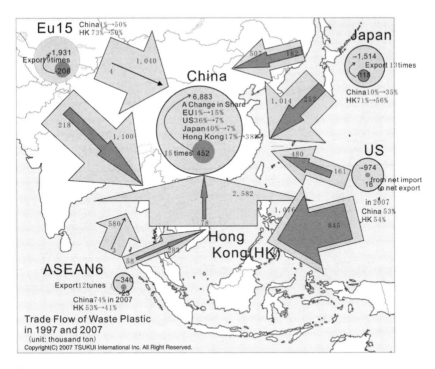

Notes:
(1) Dark gray arrows show 1997 figures and light gray arrows show 2007 figures.
(2) Percentages are export shares for exporting countries and import shares for importing countries.

Source: Author's drawing. UN Comtrade and World Trade Atlas.

Figure 2.2   Trade flows of waste plastic and differences between 1997 and 2007

paper require further separation.[6] For example, electric wire is composed of copper wire and plastic covering, and to obtain the copper scrap, the plastic cover needs to be stripped, a process that requires manual work. As another example, iron scrap from demolition debris needs to be separated from the concrete. As recyclable resources differ in shape and the manner in which they are mixed with other materials, separation work needs to be done, at least partially, by workers. Therefore, separation work is one of the key processes for recycling and as such, developing countries that are more competitive in labor-intensive work are competitive in the recycling processes. Consequently, recyclable resources are recovered at lower cost in these countries.

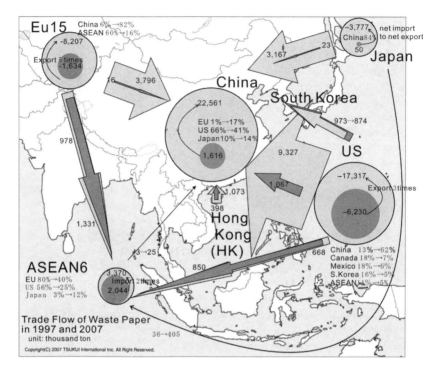

Notes:
(1) Dark gray arrows show 1997 figures and light gray arrows show 2007 figures.
(2) Percentages are export shares for exporting countries and import shares for importing countries.

Source: Author's drawing. UN Comtrade and World Trade Atlas.

Figure 2.3   Trade flows of waste paper and differences between 1997 and 2007

The second factor is the cost of recycling. Some products using recycled resources can be produced competitively because of the lower processing costs and using the resources is less expensive in developing countries. Waste plastic is a good example of this. In China, plastic recovered from waste PET bottles is used for making synthetic fiber, toys, and other plastic products. Japan does not have a competitive advantage in producing such goods from waste PET bottles. Japan exports waste plastic to China at higher prices than when exchanged in the Japanese market not only because the recycling costs of waste plastic are lower in China, but also because more products can be produced competitively from the waste there.

Third, in terms of consumption patterns, it is often suggested that lower income consumers demand lower quality and cheaper products, rather than their more expensive and higher quality counterparts. Lower quality but cheaper products produced from recycled resources are in greater demand in developing countries compared with in developed countries and therefore, greater demand for recycled resources is found in these developing countries. This pattern resulted in many recyclable resources being exported to countries with higher demand for these resources and at higher prices.

The final factor relates to environmental policy. It is often observed that developing countries have less stringent environmental regulations, in part because the governments' capacity to implement the regulations is not sufficient. As a result, many recycling industries choose to relocate to lower income countries with less stringent regulations.

All the points made above suggest that recyclable resources of lower quality or those of lower separation quality should be in greater demand in less developed countries. This speculation is backed up by empirical study. Michida et al. (2011) indicate that in the case of iron and steel scrap, imported and exported scrap differs in quality. Indeed, we observe from customs statistics that prices of traded iron and steel scrap vary significantly depending on the trading partners. For example, the unit value of iron and steel scrap (HS7204) exported from Japan to the United States was ten times higher than the iron and steel scrap exported to China in 2007. The empirical analysis of the paper shows that recyclables of lower separation quality, which are therefore priced lower, are sent to lower income countries as well as to countries with less strictly enforced environmental regulations.[7]

On the supply side, recycling legislation has been introduced in developed and middle income countries and these laws have contributed to an increase in the amount of recyclables. In Japan, recycling policies such as the Containers and Packaging Recycling Law of 2000, the Construction Waste Recycling Law of 2002, and the End-of-Life Vehicle Recycling Law of 2005 were enacted. The implementation of these laws resulted in an increase in the amount of recyclables in Japan. Figure 2.4 shows the trend in Japanese exports of certain recyclables, and it can clearly be seen that waste paper and waste plastic exports increased after 2000. In Asia, Japan has been a major supplier of many recyclables in the region. It should be noted that Japan used to be a recyclable importer and became a net exporter of some recyclables after the 1990s (Figure 2.4). With changes in domestic demand and supply of recyclables as well as changes in comparative advantage, countries may shift from being importers to exporters. This fact leads us to consider that in the long run, there may be structural changes in the current trends of recyclables trading.

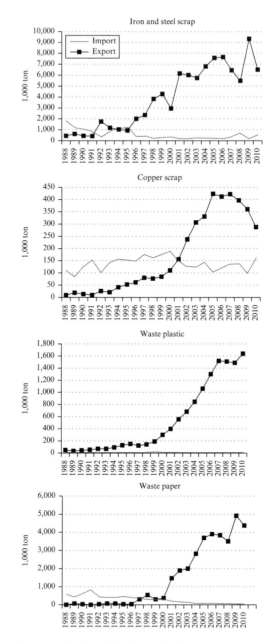

Source: Created by the author from UN Comtrade data.

Figure 2.4   Japanese export and import of recyclables from 1988 to 2010

2.3 OTHER DETERMINANTS OF RECYCLING INDUSTRY LOCATION

Many emerging Asian countries import recyclables and use recycled materials to meet domestic demand for recyclable resources. As we saw in the Chinese examples, it seems some countries are more competitive in importing certain recyclable resources than others. From the previous discussion, there are multiple factors that affect the trade flow of recyclable resources. However, the dominating factors are not the same across different recyclables and countries. Countries with lower labor costs are competitive in processing and using recyclables, and this seems to be a major factor in the exportation of recyclables to developing countries. However, labor cost is not the only determinant factor for the location for recycling industries. If income levels were the only determinant, all recyclables would be exported to lower income countries, but this is not what has been observed (Table 2.1).

Table 2.3 shows the revealed comparative advantage[8] in terms of imports to examine import competitiveness in the region for certain recyclables. Depending on the recyclables, import competitiveness differs

Table 2.3   Revealed comparative advantage in imports 2005–07 (average)

	Waste plastic	Waste paper	Waste glass	Precious metal scrap	Iron and steel scrap	Copper scrap	Aluminum scrap
Japan	0.0074	0.0550	1.6083	**1.4417**	0.3053	0.9166	0.5422
China	**6.8482**	**6.6612**	0.8780	0.0009	1.1081	**4.8705**	**3.1107**
Hong Kong	**10.2652**	0.0692	0.0496	0.9454	0.0978	0.8121	0.1123
S. Korea	0.0868	1.4095	**1.9647**	1.2322	**3.1366**	**2.8940**	**2.8462**
Taiwan	0.4753	1.0560	0.1490	0.0444	**3.4113**	2.6469	1.5687
Singapore	0.0579	0.0435	0.6358	1.1689	0.1828	0.2405	0.0513
Thailand	0.0190	2.0862	1.1349	0.0112	1.4447	0.1945	0.8236
Malaysia	0.5157	0.4629	**4.2813**	0.2772	2.3781	0.5232	0.7623
Indonesia	0.0490	**9.1672**	0.6177	0.0001	1.8305	0.1341	0.8640
Philippines	0.0676	1.0598	0.3828	0.0001	0.0391	0.2087	0.0120
Vietnam	0.9735	1.4793	0.2911	0.0055	1.3191	0.2069	0.1112

Notes:
Let M denote the import value. For good k of country i, the index is calculated as $\frac{M_k^i}{M_K^i}$ where K denotes all goods and W denotes the world. $\frac{M_k^i/M_K^i}{M_k^W/M_K^W}$ is the world import share of good k in total world imports.
The first and second highest indices are shown in bold type for each recyclable.

Source: Author's calculations using UN ComTrade data.

among countries. China and Hong Kong have higher competitiveness in importing waste paper and waste plastic, as is shown in Figures 2.2 and 2.3, but with regard to aluminum and copper scrap, South Korea is relatively competitive. For iron and steel scrap, South Korea and Taiwan have a greater comparative advantage. Therefore, determinants other than income levels also affect location of recycling industries.

The market size that determines the demand for recycled materials seems to be an important factor for waste plastic or waste paper. If countries have larger markets for the recycled materials, firms have an incentive to import and process recyclables as they can easily sell the recycled materials domestically. For example, China exports electrical appliance parts, as well as various kinds of final goods, all over the world, and cardboard is needed for packing and shipping these goods. Waste paper is used to produce cardboard and thus demand for cardboard creates a large market for recycled paper in China.

Know-how and technology play another important role. As shown in Table 2.3, Japan and South Korea have higher competitiveness in importing precious metal scrap. This is because of the higher level of technology required to recover the precious metals. Also the value of recycled metals is higher and the higher value can help pay for the recycling costs in these countries. Rare metal or rare earth resources could be categorized here as resources that require higher technology to recover, and at the same time these resources are indispensable in various industrial production processes. However, the recovery costs for these metals remain expensive because waste electronics and electrical scrap contain only a small amount of the metal, which pushes up the costs of recovery. Therefore, the recovery of rare metal and rare earth resources has not yet achieved a competitive edge.

Path dependency also plays a role. An example is the Indonesian import of waste paper, an area in which Indonesia has high import competitiveness. Indonesia has vast forest resources and the country has large paper mills to utilize the country's abundant wood pulp. Partly because of a lack of raw materials, Indonesia imports waste paper to feed the mills. This background explains the higher competitiveness of waste paper imports in Indonesia. Another example for path dependency is waste glass. A large amount of cathode ray tube (CRT) monitors have become waste as the demand for TV sets has shifted from products with CRTs to those with flat panels. CRT monitors contain lead, so the best method of recycling them is to reproduce the CRT monitors. However, there are few factories that still produce CRT monitors as product demand has decreased globally, but these factories are the only places that can properly recycle the increasing number of waste CRT monitors. In the Asian region, only

South Korea and Malaysia have factories large enough to recycle imported waste CRT monitors and produce new products. Waste glass is therefore shipped to South Korea and Malaysia. This trend is also suggested by the figures given in Table 2.3.

In China, the government has long encouraged the use of recycled material and the country is experienced in utilizing the materials (Yamaguchi 2003). This background enables the country to accommodate large-scale recycling. In Vietnam, villages specialize in dealing with specific recycled materials, and residents share the knowledge needed to perform the recycling (Sakata 2007 and 2009).

2.4 TRADE FLOW WITH ENVIRONMENT AND TRADE REGULATION

As the trade of recyclables has expanded, there are signs that the trade activities have resulted in some degree of environmental problems in Asian countries. Two types of problems have been identified. The first concerns problems related to pollution emitted from the recycling processes conducted in developing countries. The informal sectors without appropriate pollution treatment facilities are often involved in the recycling activities. Thus, residue after separation of the recyclables ends up as a waste and may be dumped illegally. The second problem is the illegal trade of the recyclables. The Basel Convention set up international rules to prevent hazardous waste dumping in developing countries; however, the trade of recyclables is not restricted by international treaties as recyclables are considered as goods or resources. However, some cases of illegal trade have caused problems. For example, hazardous waste that should be controlled under the Basel Convention has been mixed with recyclables or disguised as recyclables and exported to developing countries where monitoring at the borders is not strictly enforced.

Ensuring the sustainability of the recyclables trade is important since it has been found to be beneficial for both importing and exporting countries. Increasing resource productivity through the trade of recyclables across borders, and, at the same, utilizing the recycled resources in a sustainable manner has become an important and challenging policy issue. Both exporting and importing countries have been tackling the related environmental problems by adjusting environmental and trade regulations. Some policies have been introduced to maximize the benefits of recyclable resource utilization and minimize the negative externalities by restricting the recyclables that cause damage to the environment. The Chinese government's reaction to these problems is one example: it has introduced regulations on the importation of recyclables. According to Yoshida (2010), to combat the environmental problems, the Chinese government requires recyclable importers to obtain permits from the government and show that they are able to control pollution appropriately. Moreover, the government issues a list of importable recyclables, which are divided into two categories: recyclables of lower environmental impact and those of higher environmental impact. The listed recyclables are regulated according to pollution risks. The list has been updated every year and underwent major revisions in 2001, 2002, and 2003. Some goods were added or re-categorized in each revision.

In 2009, Indonesia issued the Decree of Minister of Trade of the Republic of Indonesia Regulation No. 39/M-DAG/PER/9/2009, which regulates the importation of recyclables in a similar way as in China. The Decree states that certain industries in the country still use non-hazardous and toxic waste (non-B3 waste) as raw materials for production, and that sufficient amounts of non-B3 waste as raw materials needed for the certain industries could not be obtained solely from domestic sources, requiring additional supply from overseas sources. Further, the Decree stated that the provision of non-B3 waste as raw materials from overseas sources must continue to observe environmental protection programs in the country, thus making it necessary to control and restrict the imports. In the Decree, the Indonesian government made it clear that it is necessary to import recyclables as resources and set import regulations to protect the environment. The Decree requires importers to register and pass technical verification, and it provides a list of importable non-hazardous and toxic waste.

Stricter regulation is necessary to prevent pollution. However, if regulations are too stringent, this acts as an impediment to the trade and utilization of recyclable resources. Therefore, there have also been some cases of deregulation. For example, technological development makes some recycling processes less polluting. This allows governments to lift the import ban on these recyclables or loosen regulations to improve utilization. In China, the government had imposed regulation requiring that waste plastic bottles be crushed into pellets or flakes before importing. Later, the regulation was changed to allow non-crushed plastic bottles to be imported. This regulatory change was made because new factories able to process the PET bottles started operations.

Setting efficient regulation is not an easy task. As we can see from the revision of the recyclables list in China, regulations and border control procedures need to be updated constantly as the smuggling of hazardous waste disguised as recyclables has become rampant, and control methods need to be revised to ensure regulatory effectiveness. Also, the regulations need to reflect the technology and capacity of the recycling industry to ensure the resources can be used efficiently as well as in an environmentally friendly manner. It is a challenge for importing countries to balance the benefits and the negative externalities of imported recyclables by setting appropriate regulations.

Let us look now at an example of the reactions of exporting countries to some cases of illegal trade. CRT monitors and lead batteries are often found in illegal exports. These goods are claimed to be second-hand goods, but their actual condition suggests that they are waste. There are 13 cases of ship-backs of CRTs and lead batteries from Hong Kong to Japan from 2005 to 2007, according to Japan's Ministry of the Environment website. There are similar cases in trade between Hong Kong and the United States, with dozens of ship-back cases reported in the shipment of waste lead batteries from Hong Kong to the United States. These cases include those in which the shipment was reported as scrap when exported, but was considered as waste lead battery by the Chinese Customs Office. In another case in 2009, CRT monitors were illegally shipped to Semarang, Central Java in Indonesia from Massachusetts in the United States. As CRT monitors contain lead, it was suggested that they should have been subject to the rules set by the Basel Convention. Exporting countries have taken measures to prevent such illegal exports. For example, the Japanese Ministry of Economy, Trade and Industry (METI) issues guidelines and criteria regarding second-hand CRTs,[9] and the US Environmental Protection Agency (EPA) imposes penalties for illegally shipping CRTs.[10]

2.5 DISCUSSION

Regulatory changes in countries importing recyclables affect the exporting countries' waste management through changes in trade flow. For example, Japan ships a large amount of recyclables to neighboring countries, as domestic recycling is not economically viable because of the high cost of resource recovery and low demand for recycled resources. The trade of recyclables helps Japan to reduce waste as some recyclables might be dumped as waste domestically, which shortens the lifetime of final disposal sites. Moreover, if recyclables cannot be exported, some of them will definitely become waste because the domestic recycling industry does not have the capacity to handle all the recyclables produced domestically. If recyclables such as metal scraps, waste paper, or waste plastic are not exported for utilization, the prices could drop from oversupply in the domestic market.

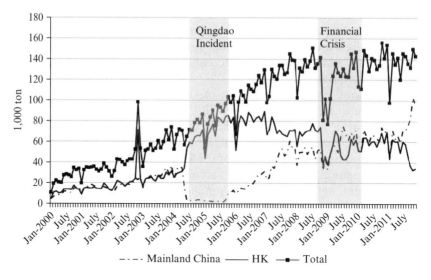

Source: Created by the author from World Trade Atlas data.

Figure 2.5   Waste plastic exports from Japan to China

The large trade flow of recyclables might pose risks to trading countries because some imported recyclables are concentrated in certain countries for processing, and a regulatory change in a single country could affect the world trade flow of these recyclables. The processing of some recyclables such as waste paper and waste plastic has been concentrated in China. Waste glass from CRT monitors has been imported and recycled only in Malaysia and South Korea. As such, many countries both inside and outside the region rely on a small number of countries for the recycling of certain materials. The geographic concentration of some recycling industries poses risks to the exporting countries because restrictions on trade from the sudden introduction of import controls will result in a stockpile of the recyclables in these countries. A clear example of this occurred in March 2004 when the Chinese government claimed that 4,000 tons of waste plastic was exported illegally from Japan to Qingdao, China. Subsequently, China banned all imports of waste plastic from Japan from May to September of the same year. This regulation decreased waste plastic exports from Japan to mainland China during the period, as shown in Figure 2.5 (see Yoshida 2010, and Chapter 3 for details about the Qingdao Incident). Many Japanese export firms were later able to redirect their shipments to Hong Kong (Kojima and Yoshida 2005). However, for some time, waste plastic collected for recycling piled up domestically as the recycling capacity within the country was far from sufficient to process all that was collected. The Qingdao Incident was a lesson to Japan that a regulatory change in a single country can have a significant impact on the trade of recyclables.

The procurement of rare metals is becoming a hot issue as China is the largest single source of these metals and export controls of rare metals affect production in industries such as car manufacture and electronics. The control of risk in procurement for inputs is very important. Similar risks exist in the trade flow of recyclables. Although the concentration of recycling activities in certain countries is often one economic consequence of the various factors discussed, it is also important to consider risk diversification in terms of location.

To ensure a sustainable trade of recyclables in Asia, two policy recommendations are suggested. The first concerns the prevention of environmental pollution related to imported recyclables. As we saw, the countries that import recyclables can adjust import regulations as well as domestic environmental pollution controls by, for example, allowing only licensed firms to import and process the recyclables. Exporting countries may also try to tighten monitoring for illegal exports of hazardous wastes mixed with recyclables at their ports. The big challenge for both importing and exporting countries is to control smuggling and the activities of the informal sectors. Policies need to be implemented to formalize the informal sectors (see Hosoda 2008, for a discussion) to reduce the uncertainty related to these illegal and informal flows. The regional coordination of the extended producer responsibility (EPR) and international standards are also important policy steps, as discussed in Hotta et al. (2008).

Improving the quality of traded recyclables is another way to prevent pollution in importing countries. The quality of recyclables can be improved by having better separation quality before exporting. For example, currently iron scrap is often exported as a mixture with other types of wastes such as plastics and other metals. These other types of wastes shift waste to the importing countries. If recyclables are separated better before exporting, residue and waste received with the iron scrap is reduced in importing countries.

Improvement of recyclable quality has been pursued from various perspectives in the European Union as well. An example can be seen in the RoHS/WEEE directives.[11] The RoHS directive took effect on July 2006, and this directive restricts the use of six hazardous substances (lead, mercury, cadmium, hexavalent chromium, polybrominated biphenyls (PBB), and polybrominated diphenyl ether (PBDDE)) in various types of electrical and electronic (EE) products. If one of the component parts exceeds the specified limit, the whole product fails to meet the regulation. The

RoHS directive is closely linked with the Waste Electrical and Electronic Equipment (WEEE) directive which sets collection and recycling targets for EE products and is part of a legislative initiative to combat problems associated with the increase in e-waste and related environmental contamination. The intention of the two directives together is to manage the use of chemicals in EE products throughout their entire life cycle. In order to prevent pollution such as soil contamination from mercury, lead, and other hazardous materials leaking from disposed e-waste, and to prevent hazardous substances from entering the production chain after recycling, the European Union decided to restrict the use of such chemicals at the source in EE products when they are produced. As the RoHS directive also affects other parts of the world, some plants outside the region have begun using only waste plastics without the restricted substances as regulated in the RoHS directive for recyclable resources. Thus, the European Union directives have resulted in better quality and safer recyclables both within and outside the region (see Michida 2012).

The second suggestion relates to the increase in recycling activities in developed countries and aims to find and create a more appropriate division of labor among Asian countries. There is room for developed countries to recycle more within their borders. Debates have begun in Japan, for example, over the utilization of rare metals and other high cost metals. Urban mining—the extraction of the various metals contained in electric or construction scrap—has attracted much attention in policy debates as waste metal recycling could improve resource efficiency. Recent price increases for the natural resources as a result of increased demand in emerging countries has led to higher demand for scrap as a substitute. Generally speaking, scrap is exported from developed countries, where it is generated in relatively large volumes, to developing countries with high resource demand.

In resource-poor developed countries such as Japan, an increase in scrap exports has given rise to a debate as to how to conduct efficient urban mining domestically. This discussion is often framed around how the export of recyclable waste is regarded as a leakage of resources by the exporting country. In fact, although a large volume of iron and steel scrap is exported to developing countries, some scrap is exported to developed countries. This implies that there might be some room for developed countries to increase domestic recycling if the appropriate quality of recyclables is available. Some recyclables, such as precious metals or hazardous waste, require more sophisticated technology to process, so for these items, developed countries have an advantage. For other recyclables, if well separated and not requiring labor-intensive work to use, developed countries could also possibly do the recycling domestically. Policies should be implemented to increase the supply of recyclables that corresponds with the technology in developed countries.

NOTES

1. In this chapter, terms such as trade and environmental policies or regulations are used as general terms. These refer to legally binding regulations, non-binding regulations, standards, or guidelines depending on the context.
2. With regard to the statistics of China, this chapter refers to mainland China and Hong Kong separately when using statistics on the separate areas, references of which are stated in the tables and figures. In the main text, we generally refer to China as mainland China for simplification.
3. HS statistics codes for recyclables are as follows: Waste paper (4707), waste plastic (3915), waste glass (7001), aluminum (7602), iron and steel scrap (7204), copper scrap (7404), and precious metals (7112).
4. World total import refers to the world total import recorded in the UN Comtrade database. Asia refers to China including Hong Kong, Japan, the Republic of Korea (South Korea), Taiwan, and ASEAN 5 (Indonesia, Malaysia, Philippines, Thailand, and Vietnam).
5. The author calculated these figures using World Trade Atlas data.
6. Recyclables differ from one another even for the same resource in separation quality. More homogeneous recyclables are generated from emitters such as large factories that produce homogeneous residue or cut metals.
7. Michida et al. (2011) also show that the quality of exported scrap rises with income levels. This finding suggests that as income rises, further separation of recyclables is conducted and higher quality recyclables are produced.
8. Balassa (1965) shows the index of revealed comparative advantage for export.
9. See Japan METI's website, http://www.meti.go.jp/policy/recycle/main/admin_info/law/10/pdf/exp_tv.pdf for the export rule on CRTs for Japanese exporters.
10. See US Environmental Protection Agency website, http://www.epa.gov/osw/hazard/recycling/electron/#crts for the US export rule on CRTs.
11. RoHS directive (Directive 2002/95/EC of the European Parliament and of the Council of 27 January 2003) is the Directive on the restriction of the use of certain hazardous substances in electrical and electronic equipment and entered into force in July 2006. WEEE is the Waste Electrical and Electronic Equipment Directive (Directive 2002/96/EC of the European Parliament and of the Council of 27 January 2003), which aims to impose the responsibility for the disposal of waste electrical and electronic equipment on the manufacturers of such equipment in an ecological manner.

REFERENCES

Balassa, Bela (1965), 'Traded liberalization and "revealed" comparative advantage', *The Manchester School of Economic and Social Studies*, **33**, 99–123.

Copeland, Brian R. (1991), 'International trade of waste products in the presence of illegal disposal', *Journal of Environmental Economics and Management*, **20**, 143–62.

Hotta, Yaushiko, Mark Elder, Hideyuki Mori and Makiko Tanaka (2008), 'Policy considerations for establishing an environmentally sound regional material flow in East Asia', *The Journal of Environment & Development*, **17** (1), 26–50.

Levinson, Arik (1999), 'State taxes and interstate hazardous waste shipments', *The American Economic Review*, **89** (3), 666–77.
Ley, Eduardo, Molly Macauley and Stephen W. Salant (2000), 'Restricting the trash trade', *The American Economic Review*, **90** (2), 243–6.
Ray, Amit (2008), 'Waste management in developing Asia: can trade and cooperation help?', *The Journal of Environment & Development*, **17** (1), 3–25.
Michida, Etsuyo, Cemal Atici and Michikazu Kojima (2011), 'Does quality matter in iron and steel scrap trade?', *IDE Discussion Paper Series*.

References in Japanese

Fujii Yoji (2010), 'Sekai no Saiseishigenboueki no Genjyo: Chugokuno Saiseishigenyunyu wo chushin ni' (Situation of World Recyclable Resource Trade: Importation of Recyclables in China) in *Keizaikeiei Kenkyusyo Nenpou*, Tokyo: Kantogakuin Daigaku.
Hosoda, Eiji (2008), 'Shigen Junkan Shakai -Seido Sekkei to Seisaku Tenbo' (Society with an Environmentally-Sound Material Cycle: Institutional Design and Policy Review)', Tokyo: Keio University Press Inc.
Kojima, Michikazu and Aya Yoshida (2005), 'Saiseishigen Chukohinboueki no Chukeichi toshiteno Hong Kong' (Hong Kong: Transit Center of the Trade of Recyclables, Second-hand Goods), in Michikazu Kojima (ed.), *Ajia ni Okeru Junkan Shigen Boueki* (Trade of Recyclable Resources in Asia), Chiba: Institute of Developing Economies, 69–83.
Michida, Etsuyo (2012), 'Seihin Kankyokiseiwotujita Boueki to Kankyo no Linkage: Supply-Chain wo Tsujita Tojyokoku heno Eikyo', (Trade and Environment Linkage through Product-Related Environmental Regulation: The Impact on Developing Countries through Supply Chains), Chosa Kenkyu Hokoku Sho, Institute of Developing Economies.
Sakata, Shozo (2007), 'Vetonamu no Risaikuru Mura (Recycling Villages in Vietnam)', Ajia Chiiki niokeru Risaikuru no Jittai to Kokusai Shigen Junkan no Kanri. 3R Seisaku (Recycling in Asia and Policies for Managing International Trade of Recyclable Resources and Promoting 3R), Institute of Developing Economies and National Institute for Environmental Studies, 2–13.
Sakata, Shozo (2009), 'Vetonamu Kouga Deruta Chiiki no Noson Kogyou – Risaikuru Muranotten ni Miru Shoukibo Keizaishutai no Senryaku' (Village Industry in Delta Region of Vietnam: Strategies of Small-Scale Economic Agents in Development of Recycling Villages), in Shigeo Sakata (ed.), *Henyou suru Vetonamu no Keizaishutai*, Chiba: Institute of Developing Economies and National Institute for Environmental Studies, 223–49.
Yamaguchi, Mami (2003), 'Chugoku Toshi Infomaru Sekuta niokeru Chihou Shusshinsha no Shugyou Kouzou – Pekin-shi Haihin Kaishu Gyousha no Jirei wo Chushin ni-' (Structure of Workers from Local areas in Chinese Informal Sectors of Cities), *Ajia Keizai*, **44** (12), 28–56.
Yoshida, Aya (2010), 'Chugoku ni Okeru Haikibutsu Genryo no Yunyu Kisei', in Michikazu Kojima (ed.), *Ajia ni Okeru Junkan Shigen Boueki* (Trade of Recyclable Resources in Asia), Chiba: Institute of Developing Economies, pp. 69–83.

3. Recyclable waste trade of Mainland China

Aya Yoshida

INTRODUCTION

China began importing recyclable waste in the 1990s because of a scarcity of raw materials domestically (Yoshida et al. 2005). Economic growth has continued at a high level and China requires huge volumes of resources to support its industrial demand and economic activity. In order to secure the resources needed to support the country's high rate of economic growth, China imports various forms of recyclable waste. In addition, low labor costs in China make the recycling of low-grade recyclable waste economically feasible, in comparison to other industrial nations of the world. Resource demand and cheap labor have acted as triggers for a huge wave of recyclable waste imports from overseas.

Although China is actively utilizing recyclable waste from overseas, it is also well on the way to becoming what can only be termed "the world's dumping ground." Serious environmental pollution has been generated by improper recycling. In particular, the recycling of e-waste in areas such as Guiyu is leading to environmental pollution (Yoshida 2005). The Chinese government has already taken various steps in an attempt to prevent environmental pollution, including introducing an import license system, banning imports of waste household appliances, establishing export standards, and imposing pre-shipment inspection requirements. However, cases of ship-backs and smuggling still continue.

In this chapter, using customs statistics and field surveys, we characterize the current recyclable waste trade in China and identify major trends and areas of progress that China faces in connection with recycling. In Section 3.1, the import trends in recyclable waste to China are analyzed based on detailed trade statistics. In Section 3.2, the transboundary movement of second-hand electric and electronic equipment and e-waste scrap is discussed. Section 3.3 explains the recent regulation of imported waste and trends in e-waste and waste electric and electronic equipment (EE equipment) recycling.

3.1   BASIC TRENDS IN RECYCLABLE WASTE IMPORTS TO CHINA

3.1.1   Import Trends in Recyclable Waste

According to the Chinese customs statistics, in 2010 China imported 8.0 million tons of waste plastic, 24.4 million tons of used paper, 5.8 million tons of scrap iron, 4.4 million tons of scrap copper, and 2.9 million tons of aluminum scrap. As shown in Figure 3.1, the amount of recyclable waste imported by China is growing every year. China imports recyclable waste from Japan, the European Union, United States, and its neighboring countries.

Approximately 60 percent of waste plastic was imported from Asian countries. Of the total waste plastic imported, Hong Kong accounts for 25 percent and Japan and Thailand each provide approximately 10 percent. The European Union has gradually increased its exports of waste plastic; however, Asian countries remain dominant. The amount from North America remains constant. The price of ferrous metals (steel scrap) imports fluctuates considerably over time: there was a sharp decrease in steel scrap price from 688 USD/ton in 2005 to 257 USD/ton in 2008, followed by a sharp increase to 371 USD/ton in 2009 (Figure 3.2). There are two reasons for this rapid change in the price of imported steel scrap. First, around 2006, the Chinese government enacted measures regarding small-scale electric furnaces to make steel production more efficient. Second, the price of scrap steel became very high, especially during 2007–08. Imports of copper scrap decreased sharply from Asian countries in 2009, while at the same time there was a sharp increase in steel scrap imported from Asia. Therefore, it can be assumed that metal scrap that was imported under the name of copper scrap replaced steel scrap. For aluminum scrap, the ratio from Asia, the European Union, and North America has remained constant. However, imported metal scrap from other countries and regions such as Australia and South America increased in 2005 and 2009.

3.1.2   Destination of Recyclable Waste

In order to determine the destination of secondary material flow, import data from Chinese customs statistics is analyzed by country, destination, and trade type. The destination of each secondary material from 1995 to 2007 is shown in Figure 3.3.

According to Chinese customs statistics for 2007, nearly 50 percent of imported waste plastic was concentrated in Guangdong Province

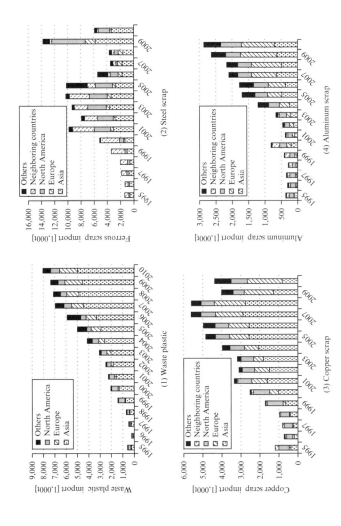

Note: Asia refers to Cambodia, Hong Kong, India, Indonesia, Japan, Macau, Malaysia, Mongolia, North Korea, Pakistan, Philippines, Singapore, South Korea, Taiwan, Thailand and Vietnam; Europe refers to EU15 (Austria, Belgium, Denmark, Finland, France, Germany, Greece, Ireland, Italy, Luxembourg, Netherlands, Portugal, Spain, Sweden, United Kingdom); neighboring countries refers to Kazakhstan, Kyrgyzstan, and Russia; North America refers to United States and Canada.

Source: China Customs Statistics.

Figure 3.1   Import volumes of recyclable waste by region (1995–2010)

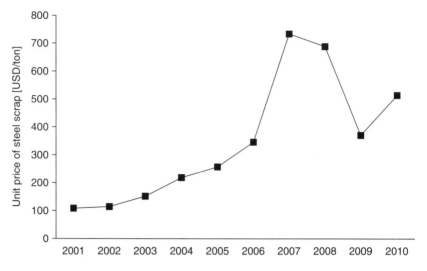

Source: China Customs Statistics.

Figure 3.2   Unit price of steel scrap (2001–10)

and approximately 30 percent went to the Huadong area (Shanghai, Zhejiang, Jiangsu, Shandong, Jiangxi, and Anhui provinces). Guangdong Province, which is well known for consignment manufacturing (processing and assembling, processing with imported materials, etc.), imports waste plastic through Hong Kong and then exports 100 percent of its plastic products to other countries. Inputs are imported free of duties and VAT (value added tax). The amount of plastic scrap imported into the Huadong area as a percent of total imports has expanded from 9 percent to 27 percent, as the market price in this area is higher due to the growing demand from the textile industry. Waste PET is mainly concentrated in Zhejiang (50 percent) and Jiangsu (30 percent) provinces because there are many textile companies that can recycle PET flakes into short polyester fiber (Figure 3.4).

Approximately 40 percent of steel scrap was imported into Guangdong Province, 20 percent to the Huadong area, and 10 percent to the Xinjiang Uyghur Autonomous Region (Xinjiang region). Approximately half of copper scrap imports were concentrated in the Huadong area and approximately 70 percent of aluminum scrap was imported to Guangdong Province. Chinese imports of copper and aluminum scrap from the neighboring countries has shifted to higher quality scrap, whereas import of steel scrap has shifted to lower quality scrap.

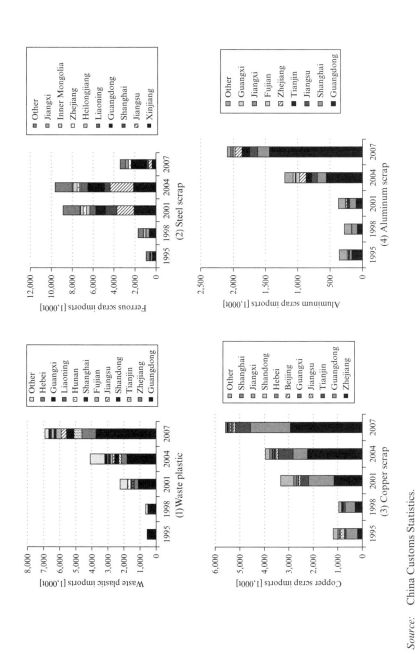

Source: China Customs Statistics.

Figure 3.3 Destination of recyclable waste imported to China (1995–2007)

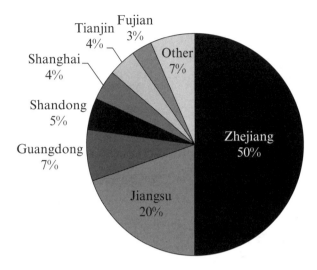

Source: China Customs Statistics.

Figure 3.4   Destination of waste PET in 2007

3.1.3   Type of Trade

Trade type can be categorized into ordinary trade, processing and assembling with supplied materials, processing with imported materials, and border trade, among others. Processing trade refers to the business activity of importing all or part of the raw materials, parts, and components from abroad in bond, and re-exporting the finished products after processing or assembly by Chinese enterprises. There are two kinds of processing trade: processing and assembling with supplied materials, and processing with imported materials. In the processing and assembling with supplied materials, raw materials and components are supplied by a foreign company and processed by a Chinese enterprise. Ownership of raw materials and components remains that of the foreign company. The Chinese company does not have to make foreign exchange payments, and is paid through the charging of a processing fee. On the other hand, in processing with imported materials, a Chinese company purchases raw materials and components, and therefore, has to make foreign currency payments. The Chinese company can export the finished products to any foreign customer after processing because the ownership of these imported commodities remains with the Chinese enterprise. Border trade means importing and exporting permitted items or goods from neighbor countries to a licensed Chinese company in the government approved border area. Import duties and VAT exemptions are applied for some items or goods.

The current distribution among these categories is different from that in the early 1990s. Ordinary trade of waste plastic has increased from 13 percent to 95 percent, whereas processing trade has decreased significantly (Figure 3.5).

In 2001, steel scrap and aluminum scrap from neighboring countries such as Russia and Kazakhstan were imported by border provinces through border trade, whereas scrap materials from other countries were imported by coastal regions through ordinary trade. Most high-quality non-ferrous scrap is imported for processing and assembling or as border trade by border provinces such as the northeast provinces and the Xinjiang region (Yoshida et al. 2005).

Border trade accounted for 60 percent of the total amount of imported steel scrap in 1998, which consisted mostly of low quality scrap imported from neighboring countries. However, since then, ordinary trade has been growing faster than border trade. Ordinary trade accounted for 97 percent of the total amount of imported steel scrap in 2007, whereas border trade made up only 0.4 percent. Steel scrap from Russia and Kazakhstan is imported by the northeast provinces and the Xinjiang region through border trade, whereas steel scrap from other countries is imported by coastal provinces such as Shanghai and Guangdong through ordinary trade.

For copper scrap and aluminum scrap, ordinary trade has been growing faster than processing and assembling, processing with imported materials, and border trade. Ordinary trade made up 99.7 percent and 79 percent of the total amount of imported copper scrap and aluminum scrap, respectively, in 2007. High quality aluminum scrap is imported for processing with imported materials or through border trade (Table 3.1).

3.1.4   Type of Enterprise

In 1995, state-owned enterprises accounted for 60 percent, 92 percent, 71 percent, and 76 percent of imports of plastic, steel, copper and aluminum scrap, respectively (Figure 3.6). Three-fourths of enterprises that import waste plastic have been changed to private enterprises. Among steel scrap importers, the numbers of Chinese-foreign cooperative enterprises, Chinese-foreign joint ventures, and private enterprises have increased while those of state-owned enterprises have decreased. While the number of state-owned enterprises is decreasing, the total numbers of enterprises were 7.34 million, 7.97 million, 0.17 million, 1.38 million, and 0.43 million in 1995, 1998, 2001, 2004, and 2007, respectively. For copper scrap, the

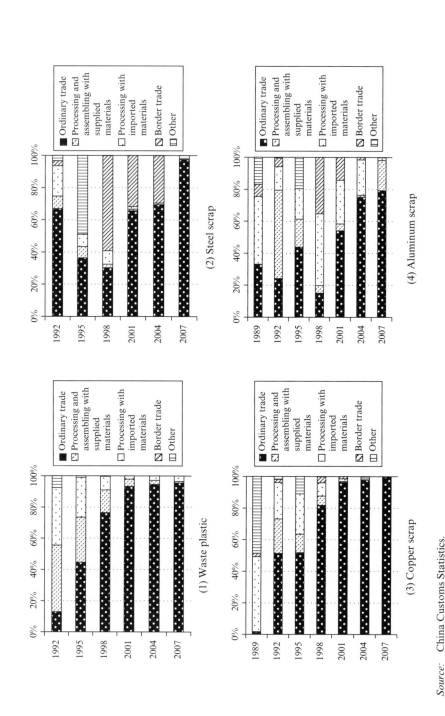

Source: China Customs Statistics.

Figure 3.5 Import of recyclable waste (by trade type)

Table 3.1   *Unit value of secondary material by trade type, 2007
                (USD/kg)*

	Plastic	Steel	Copper	Aluminum
Ordinary trade	0.46	0.75	1.13	1.02
Processing and assembling with supplied materials	0.45	0	0	1.75
Processing with imported materials	0.67	3.26	5.69	1.82
Border trade	0.35	0.22	3.58	2.05
Other	0	0.35	0	0
Total	0.46	0.74	1.14	1.17

Source: China Customs Statistics.

share of exclusively foreign-owned enterprises and private enterprises has increased; for aluminum scrap, the share of the Chinese-foreign joint ventures, exclusively foreign-owned enterprises, and private enterprises has increased.

Basic trends in recyclable waste imports to China from 1995 to 2010 show that as recyclable waste imports increase, the shares of ordinary trade and private enterprises have increased. The majority of recyclable waste imported to China was recycled in coastal areas. At the beginning, the processing trade business of recyclable waste provided job opportunities. As domestic demand for materials increases in China, ordinary trade becomes dominant and it has supplemented the resource necessary for rapid economic development in coastal areas of China.

3.2   TRANSBOUNDARY MOVEMENT OF SECOND-HAND ELECTRIC AND ELECTRONIC EQUIPMENT AND E-WASTE SCRAP TO CHINA

3.2.1   Second-hand Electric and Electronic Equipment Imports to China

Many home appliances at the end of their useful life that are discarded in Japan are reused in Southeast Asia (Yoshida and Terazono 2010). Material flow of second-hand EE equipment around China and Vietnam was surveyed by Shinkuma and Huong (2009), who found that second-hand EE equipment and e-waste scrap are still being imported to China through Hong Kong, which is a free port, to Guangdong Province, even after the import ban by the Chinese government took effect.

The government of Hong Kong has also taken various measures for

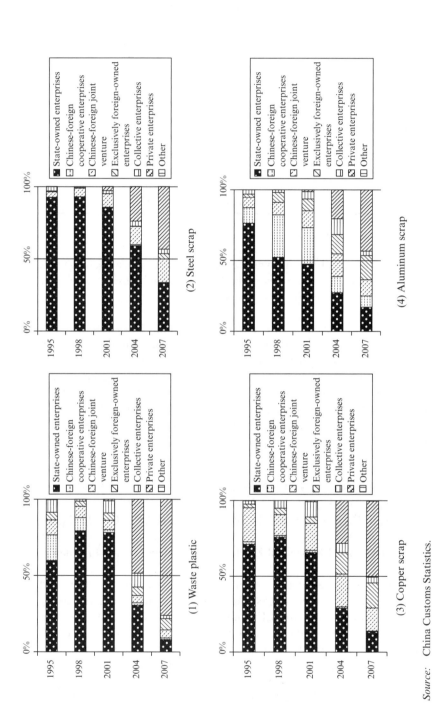

Figure 3.6   Enterprise type of recyclable waste importers (1995–2007)

(1) Waste plastic

(2) Steel scrap

(3) Copper scrap

(4) Aluminum scrap

Source: China Customs Statistics.

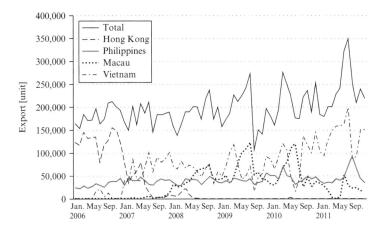

Source: Trade Statistics of Japan.

Figure 3.7   Destination of second-hand CRT televisions exported from Japan

tackling EE equipment waste imports to the region. In April 2006, Hong Kong strengthened its border controls on CRT televisions and monitors by discouraging the import of any unit more than five years from the date of manufacture.[1] This measure dramatically decreased the amount of second-hand CRT televisions exported from Japan to Hong Kong. Instead, these CRT televisions were exported to Macao or Vietnam (Figure 3.7). While Vietnam also bans in law the import of second-hand EE equipment (except laptop personal computers), in practice these imports are still continuing. Indeed, more than 50,000 units of second-hand CRT televisions are imported into Vietnam each month, and it is hard to see domestic demand at this level. Most likely the imported EE equipment is ultimately transferred to mainland China through regions with looser import controls.

Limited data are available for understanding the flow of second-hand EE equipment to China. However, detailed Chinese customs data on CRT monitors and televisions by trade type and transport method were analyzed. As a result, it is estimated that approximately 750,000 units of CRT monitors and 870,000 units of CRT televisions were imported by the river or sea to Guangxi Province, while nearly 200,000 CRT monitors and televisions were imported to Guangdong Province by land. Since these imports were classified as processing with imported materials, it can be assumed that the imported items were processed (refurbished) then exported again to other countries. In 2007, approximately 668,000 units of

CRT monitors and 900,000 units of CRT televisions were exported under the trade type processing with imported materials.

3.2.2   Imports of E-waste Parts and Scrap to China

Although the Chinese government bans the import of e-waste parts and scrap, there are many other Asian countries that export e-waste to China.

Our field surveys in Vietnam, the Philippines, and China in 2009 and 2010 show that materials recovered and sorted from e-waste such as plastics, printed circuit boards, and CRT television tubes and glass are often exported to China (NIES et al. 2010). In Vietnam, one large dealer in Ho Chi Minh City exports 100 tons of waste plastic and 60–70 tons of printed circuit boards to China each month. An interview with the company revealed that these materials are transported by ship and land to Mong Cai, the border region of Vietnam and China. There are seven large Chinese brokers in Ho Chi Minh City, so many types of materials for export are concentrated there, including printed circuit boards from Cambodia and central Vietnam. Also, many people in recycling villages in north Vietnam reported in a field visit that a variety of materials are exported to China, including plastics, printed circuit boards, and CRT tubes and glass. However, according to Chinese customs statistics, only 590 tons of waste plastic were imported from Vietnam in 2009. There is a mismatch between trade statistics and reality.

Exports of e-waste to China happen for several reasons. Printed circuit boards are exported there because the country has techniques to reuse integrated circuit parts for manufacturing other new EE equipment such as electronic toys. Also, unlike most other Southeast Asian countries, it has the technology and facilities to recover copper from printed circuit boards.

The trade volume of second-hand EE equipment and scrap flowing across the borders of China are difficult to track because some are imported without passing through customs.

3.3   CURRENT TRENDS IN RECYCLING IN CHINA

3.3.1   Recent Regulation on Importing Recyclable Waste

Since 1995, the Chinese government has implemented its own policies for the promotion and rationalization of waste import. In order to prevent environmental pollution caused by improper recycling, an import permit system and pre-shipment inspections were introduced in the late 1990s. The government also centralized the licensed mixed metal scrap recycling companies in one area and developed several industrial parks in the country to supervise recycling facilities. Overseas supplier regulation and domestic consignee registration were implemented in 2003 and 2007, respectively. The most recent regulation, Measures on Administration of Waste Import, came into force on August 1, 2011.[2] This comprehensive measure linked and extended several previous rules and transitional provisions, including the Interim Provision on the Administration of Environmental Protection in the Importation of Waste Materials in March 1996, Management Rules for Pre-shipment Inspections of Imported Wastes in September 1996, and the Notice relating to the Temporary Registration of Foreign Suppliers of Waste Materials in December 2003. Moreover, a closed-loop management system has been built for waste import in China, which has helped to reduce rejected cargos and ship-backs of improper cargos and to regulate illicit traders and recyclers.

The Chinese government regulates the type of waste that can be imported. It lists the solid waste that can be used as raw materials in China and has amended the list several times according to the changing circumstances in the country. Under the previous environmental standard (GB16487), waste plastics needed to be crushed and washed with water to prevent importation of contaminated plastics. However, compressed PET drinking bottles were added to the new list in 2009, so baled PET bottles can now be imported. This change is due to the establishment of a large-scale advanced PET bottle crushing and washing facility in the country.

3.3.2 Recent Situation in Guiyu, A Well-known E-waste Recycling Town

In rural agricultural and fishing communities in China, recycling is performed by smaller private businesses or as cottage industries by individual farmers. As a result, the inappropriate disposal of e-waste is resulting in the spread of environmental pollution.

In 2002, the Basel Action Network (BAN) and the Silicon Valley Toxics Coalition (SVTC) issued a report on the recycling industry in China, especially in Guiyu in Guangdong Province, entitled "Exporting Harm: The High-Tech Trashing of Asia." This report stated that residues from the treatment of e-waste exported by industrial nations were being illegally dumped in China and that unprocessed contaminated water was being released into rivers.

Guiyu is a town with a population of 139,000 people. Annually, nearly 1 million tons of printed circuit boards and 3 million e-waste parts are treated in Guiyu. In 2008, the sales amount from the collection and treatment of e-waste and waste plastics was 2 billion CNY (304 million USD,

1 USD = 6.58 CNY). Most of the e-waste is scrap from mobile phones and personal computers, and rarely from television sets and refrigerators. According to a report published by the local government in Guiyu on December 18, 2009, e-waste dismantling and processing was conducted in 21 villages by 300 enterprises, 5,500 individual enterprises, and 60,000 workers. However, because of the economic crisis, the number of migrant workers was reduced by two-thirds (40,000). On average, workers earn 50 CNY (8 USD) per day.

Concerns have been expressed about air, water, and soil pollution and about the damage to human health since waste in Guiyu has been recycled using primitive methods. These methods include simple manual sorting and dismantling work, open burning of plastic-coated wire, burning of soldered circuit boards, and strong acid waste treatment (Yoshida 2005). Many scientific studies have been carried out to examine the contamination in Guiyu and have revealed that the environment in Guiyu is highly contaminated. For example, the concentration of lead in the surface water of Guiyu was eight times higher than the threshold of environmental quality standards defined by the US Environmental Protection Agency. River sediments in Guiyu display lead concentrations between 28.6 and 590 mg/kg (Wong et al. 2007). The concentrations of persistent organic pollutants (POPs) and heavy metals/metalloid detected in the Guiyu air were high when compared with those from other cities. The pollution status of polychlorinated dibenzo-p-dioxins and dibenzofurans (PCDD/Fs) in a mixture of burnt residue and soil in Guiyu was 967,500 pg g^{-1} (Zheng, G.J. et al. 2008). Dust collected at the recycling workshops for printed circuit boards in Guiyu showed statistically higher concentrations of lead than non e-waste sites (Leung et al. 2008). Zheng, L.K. et al. (2008) reported on the environmental contamination and its effect on children in Guiyu. Studies of cord blood in newborns in Guiyu indicated that lead contamination from informal e-waste processing has threatened the health of residents, especially children and infants, living near e-waste recycling areas (Xu et al. 2012).

The Chinese government tightened import regulations in 2000 and 2002. The regulation banned import of waste EE equipment and its parts or components. However, e-waste continues to be brought into Guiyu because the area has already developed and gained a reputation as an industry cluster for recyclable materials domestically as well as internationally.

Our field survey in December 2009 confirmed that several companies operating with the permission of the local government collected integrated circuits and other parts from printed circuit boards in waste mobile phones and personal computers. The collected parts were supplied to toy manufacturing factories. Also, an established cottage industry that used acid baths to recover precious metals along the riverside was closed and removed by the local government. In its place, large-scale copper smelting facilities were established, which represented a great improvement in the infrastructure for e-waste recycling.

3.3.3 Old-for-new Policy to Promote Waste EE Equipment Collection and Recycling in China

Domestic e-waste has also rapidly become a growing problem in China as domestic generation has increased significantly. The Regulation for the Administration of the Recovery and Disposal of Waste Electric and Electronic Products was issued by the State Council and entered into law on January 1, 2011. The regulation targets five items (television sets, refrigerators, washing machines, air conditioners, and personal computers) and outlines the responsibilities and duties of producers, retailers, consumers, and recycling companies. Producers and retailers of EE equipment and after-service organizations are responsible for collecting used EE equipment, and central government issues qualification permits to e-waste recycling companies, which are approved by the municipal-level environmental protection department. Based on the concept of extended producer responsibilities (EPR), central government established a fund for e-waste disposal. Manufacturers of EE equipment and consignees for imported EE equipment are required to contribute to the fund.

To prepare for the enforcement of the regulation, central government started an "old-for-new policy" in June 2009 to encourage the replacement of old EE equipment with new EE equipment in nine selected provinces and cities (Beijing, Tianjin, Shanghai, Jiangsu, Zhejiang, Shandong, Guangdong, Fuzhou, and Changsha). The first phase lasted through May 31, 2010; however, the government expanded the program to an additional 19 provinces and cities, to cover a total of 28 provinces and cities, and extended the policy through December 31, 2011. Consumers can receive a subsidy of 10 percent off the sales price of new appliances if they provide their old appliances to collectors or retailers. The maximum subsidies are 400 CNY for televisions, 300 CNY for refrigerators, 250 CNY for washing machines, 350 CNY for air conditioners, and 400 CNY for personal computers. In addition, subsidies are also provided to designated collectors and licensed recycling companies for the collection and recycling of materials, respectively. However, these two subsidies are much smaller than those provided to consumers. The collection subsidy is based on the transportation distance and ranges from 30 to 60 CNY/km; the material recycling subsidies are 15 CNY for televisions, 20 CNY for refrigerators, 5 CNY for washing machines, 15 CNY for personal computers, and nothing for air conditioners. The "old-for-new policy" has two objectives: (1) address the economic crisis and stimulate domestic consumption, and (2) explore practical collection channels and funding mechanisms for regulation. By May 23, 2010, toward the end of the first year of the program, 14 million units of e-waste had been collected and 13 million units of new appliances had been sold, with a sales value of more than 50 billion CNY (7.9 billion USD), according to data from the Chinese Ministry of Commerce. By November 2011, about two and a half years into the program, the sales volume of home appliances under the government subsidy scheme surpassed 81 million units and the home appliance replacement program stimulated more than 300 billion CNY (47.6 billion USD) in consumer spending. Before the replacement program started, recycling companies could not collect enough e-waste for their operations. After the program, the amount of old appliances collected increased more than ten-fold.

However, some recyclers reported problems with storage space, as well as a lack of standards and criteria for proper storage, dismantling, and disposal of the used appliances. Because this system allows consumers to submit any type of product in the five categories of appliances, a large amount of CRT televisions were collected at recycling facilities. Since many black and white CRT televisions were also collected, it can be assumed that some consumers discarded old appliances that they obtained from different users, including by purchasing very old appliances from the market.

Even though some problems clearly exist, this replacement policy was highly appreciated by consumers, retailer shops, recyclers, and many other stakeholders. It stimulated not only consumption, but also proper domestic waste EE equipment collection and recycling.

The success of this policy also might change Chinese recyclable waste import policy in the future. Despite the fact that there are appropriate recycling facilities for imported waste, the importation of waste EE equipment to China is banned because of the lack of sufficient facilities for domestic waste EE equipment and the environmental pollution caused by informal recycling in Guiyu. As the domestic waste EE equipment recycling business expands in China, it is possible that central government could lift the ban on the import of waste EE equipment, as happened previously in the case of PET bottles.

3.4   CONCLUSION

Imports of recyclable waste started in China in the 1980s. In the early 1990s, "waste" imported under the name of recyclables was causing serious problems in China. Since 1995 the Chinese government has controlled waste imports by enacting domestic legislation. The import of waste copper, steel, and aluminum has supplemented the resources necessary for industrial development, including those needed by the recycling industries in coastal areas of China. It also drives local economic development and provides ample job opportunities. However, there are still many problems to be resolved, including weak supervision of the transboundary movement of second-hand EE equipment and e-waste scrap because most of these items are traded without passing through customs or are passed though under a different name of goods.

Small-scale enterprises, low technological level, and secondary pollution became serious problems, but there have been some improvements in recycling technologies and the domestic collection systems. For example, some improvements in the technological level were observed in an e-waste recycling area similar to Guiyu, and increasing amounts of e-waste are being collected by appropriate e-waste recycling facilities under the replacement programs. Because of the rapid growth of the domestic waste EE equipment recycling industry, China could change its regulatory policies on recyclable waste imports in the future.

NOTES

1. A similar restriction based on the age of second-hand electronics is applied in Thailand, Malaysia, and other countries. For example, Thailand restricts the import of second-hand EE equipment that is more than three years from the date of manufacture.
2. Measures on the Administration of Waste Import (in Chinese), http://www.mep.gov.cn/gkml/hbb/bl/201105/t20110520_210978.htm.

REFERENCES

Leung A.O., N.S. Duzgoren-Aydin, K.C. Cheung and M.H. Wong (2008), 'Heavy metals concentrations of surface dust from e-waste recycling and its human health implications in southeast China', *Environ Sci Technol*, **42**, 2674–80.
National Institute for Environmental Studies (NIES), Institute of Developing Economies, Japan External Trade Organization, Kyoto University (2010), 'Classification of e-waste recycling technology in Asian developing countries', Research Report of the Research Project Financed by Ministry of the Environment, Japan during FY2009 (in Japanese).
Shinkuma, T. and N.T.M. Huong (2009), 'The flow of e-waste material in the Asian region and a reconsideration of international trade policies on E-waste', *Environ Impact Assess Rev*, **29**, 25–31.
Wong, C.S.C, S.C. Wu, N.S. Duzgoren-Aydin, A. Aydin and M.H. Wong (2007),

'Trace metal contamination of sediments in an e-waste processing village in China', *Environmental Pollution*, **145**, 434–42.

Xu, X., H. Yang, A. Chen, Y. Zhou, K. Wu, J. Liu, Y. Zhang and X. Huo (2012), 'Birth outcomes related to informal e-waste recycling in Guiyu, China', *Reproductive Toxicology*, **33**, 94–8.

Yoshida, A. (2005), 'Chapter 3, China: the world's largest recyclable waste importer', in Michikazu Kojima (ed.), *International Trade of Recyclable Resources in Asia*, Institute of Developing Economies, pp. 33–52.

Yoshida, A. and A. Terazono (2010), 'Reuse of secondhand TVs exported from Japan to the Philippines', *Waste Management*, **30**, 1063–72.

Yoshida, A., A. Terazono, T. Aramaki and K. Hanaki (2005), 'Secondary materials transfer from Japan to China: destination analysis in China', *Journal of Material Cycles and Waste Management*, **7** (1), 8–15.

Zheng, G.J., A.O. Leung, J.P. Jiao and M.H. Wong (2008), 'Polychlorinated dibenzo-p-dioxins and dibenzofurans pollution in China: sources, environmental levels and potential human health impacts', *Environment International*, **34**, 1050–61.

Zheng, L.K., K.S. Wu, Y. Li, Z.L. Qi, D. Han, B. Zhang, C.W. Gu, G.J. Chen, J.X. Liu, S.J. Chen, X.J. Xu and X. Huo (2008), 'Blood lead and cadmium levels and relevant factors among children from an e-waste recycling town in China', *Environmental Research*, **108**, 15–20.

4. Import of recyclables and environmental pollution in Vietnam: a new dilemma of development

Shozo Sakata

INTRODUCTION

Rapid economic growth and industrialization started in the early 1990s in Vietnam, which coincided with the beginning of the expansion in the trade of recyclables (scrap and recyclable wastes) in global markets.[1] Vietnam, therefore, has been able to utilize imported recyclables as resources for production from the early stage of its economic development. Vietnam's subsequent trade liberalization and efforts at integrating into the global economy have widened the country's access to the global recyclables markets. Imports of recyclables are indispensable for Vietnam's industrial development as it helps to compensate for the scarcity of domestic natural resources and the high costs of imported natural resources. Without a doubt, Vietnam's industrial development has benefitted from the increasing import of recyclables.

However, increased imports of recyclables can damage the environment, as recyclables are discarded or used materials and goods.[2] Such waste, which needs to be segregated and separated to become resources, is likely exported to Vietnam with the expectation that these processes will be done more cheaply than is possible in the exporting countries. Moreover, the likelihood of illegal imports of wastes, which are more harmful to the environment, may also increase as imports expand. Increasing the import of recyclables is imperative for industrial development in many newly developing countries, and the generation of environment-related problems can be regarded as a new "dilemma of development" in countries like Vietnam that have an abundant resource of inexpensive labor.

The purpose of this chapter is to explore this new "dilemma" in Vietnam today. First, it tries to determine trends in the import of recyclables and in environmental problems occurring in Vietnam. The trade statistics indicate that the import of recyclables is on an increasing trend as Vietnam's trade liberalization has progressed, especially after its accession to the World Trade Organization (WTO) in 2007. However, it is difficult to assess quantitatively the environmental pollution caused by the import of recyclables. Therefore, the chapter instead simply acknowledges that there has been an increase in illegal imports of recyclables. Earlier studies have proven that the trade liberalization in Vietnam has exacerbated environmental pollution as pollution-intensive industries have moved to Vietnam (e.g., Pham Thai Hung et al. 2008). However, the impact from the increased amount of imported recyclables on the environment has yet to be studied.

Second, the chapter examines the development of the legal system for addressing cases of environmental pollution caused by the import of recyclables. The legal system related to environmental protection in the recycling industry in Vietnam has been studied in detail by Japanese scholars. For example, Kojima (2005) and Kojima and Yoshida (2007) have studied the controls over waste management in recycling activities (i.e., management and disposal of hazardous wastes and environmental criteria for discharged water and air) prior to the enactment of the 2005 Environmental Protection Law, and Kojima and Sakata (2008) have examined the regulations after 2005. On the other hand, studies on the import restrictions in place to prevent environmental pollution are limited to those conducted before the 2005 law took effect (see, for example, Kojima 2005, Kojima and Yoshida 2007). This chapter reviews the new regulations put in place after 2005 relating to the prevention of improper imports of the recyclable wastes.

Third, this chapter identifies existing problems with enforcing the environmental regulations for the import of recyclables. The regulations on environmental protection in Vietnam, especially after the 2005 Environmental Protection Law, are, in principle, strict enough to tackle the complex contemporary environmental problems. However, certain socio-economic as well as institutional circumstances have lessened the effects of such strict regulations.

The rest of the chapter is organized as follows. Section 4.1 outlines the trends and characteristics of the import of recyclables to Vietnam, using data from the World Trade Atlas.[3] Section 4.2 reviews the recent legal settings relating to the imports of recyclables and environmental protection. Section 4.3 introduces cases of illegal imports and examines the causes behind them.

4.1 TRENDS IN THE IMPORT OF RECYCLABLES TO VIETNAM

4.1.1 Participation in the Global Economy and Increased Import of Recyclables

Figure 4.1 shows the trend in the import of recyclables (iron scrap, aluminum scrap, waste plastic, and waste paper) to Vietnam, in volume terms, from 2000 to 2009. By the late 2000s, import volumes for all of these recyclables had become several times greater than the volumes in 2000, although they dropped sharply in 2009 as a result of the global economic slowdown. Import volumes started to increase rapidly in the mid-2000s, although the pattern of increase has not been uniform.

Other data show not only an increase in the import volume, but also a change in the sources of recyclables. Figures 4.2 and 4.3 show trends in the imports of two major imported recyclables for Vietnam, namely, iron scrap and waste paper, by exporting country.[4] Until the mid-2000s, Vietnam had relied on Japan for most imports of iron scrap and about half of the imports of waste paper. However, from the mid-2000s, new providers, notably the United States, entered the Vietnamese market.[5]

The main factor behind such change was Vietnam's move toward trade liberalization and participation in the global economy. In 1995, Vietnam joined ASEAN, which in the 2000s had concluded a series of free trade agreements with such countries as China, South Korea, Japan,[6] and India. Vietnam ratified a bilateral trade agreement with the United States in 2001. In 2007, Vietnam achieved accession to the WTO after a long series

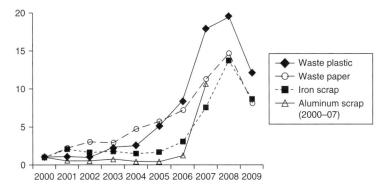

Source: World Trade Atlas.

Figure 4.1   Vietnam's import volumes of some recyclable items (2000=1)

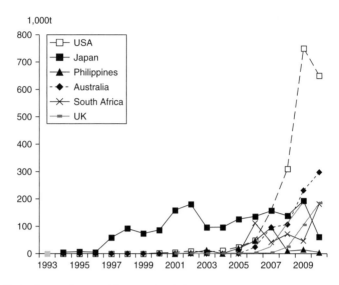

Note: Data of exporting countries are extracted.

Source: World Trade Atlas.

Figure 4.2   Vietnam's iron scrap import by country

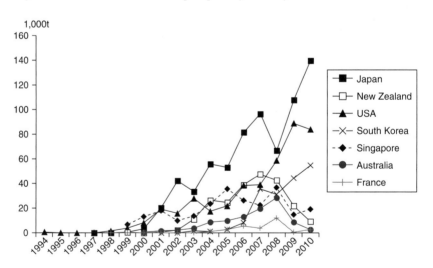

Note: Data of exporting countries are extracted.

Source: World Trade Atlas.

Figure 4.3   Vietnam's paper scrap import by country

of negotiations starting in the late 1990s. As a WTO member, the country was able to expand its international trade in terms of the number of trading partners and variety of trading goods, as well as increasing trading volume.[7]

An external factor causing the rapid inflow of recyclables to Vietnam was the trade trends in the neighboring countries. China's sudden reduction in iron scrap imports was the most noticeable case. China's import of iron scrap decreased sharply after July 2005 when the Chinese Government issued a new industrial development policy for the iron-steel sector, which aimed to abolish low-capacity, pollution-generating iron mills and to consolidate its iron production in large-scale integrated iron mills (Imai 2009). Between 2005 and 2008, China's import volume of iron scrap decreased by 65 percent. In particular, declines in iron scrap imports of 92 percent from the United States (2 million tons) and 34 percent from Australia (245,000 tons) were remarkable. The sharp increase in iron exports from the United States and Australia to Vietnam since 2006 can most likely be attributed to China's drastic decrease in iron scrap import.

4.1.2 Increase in Import of Low-quality Recyclables

Another change observed in Vietnam's import of recyclables in the 2000s was the increase in the recyclables categorized as "other." "Other" recyclables consist of mixed or low-quality, low-unit-price wastes that are more likely to lead to pollution. Figure 4.4 shows the trends in the volume of iron scrap imports (bar chart, left axis) and the proportion of "other" iron scrap (HS 720449) in the total imported volume (line chart, right axis) of Vietnam and the six other main iron scrap importers in Asia. Similarly, Figure 4.5 shows the trends in the volume of waste paper imports and the proportion of "other" waste paper (HS 470790). Comparing Vietnam with the other countries reveals a clear trend: as the volume of imported recyclables increased, the portion of the "other" categorized scrap and wastes also sharply increased in Vietnam. For other countries, although the trends in the import volumes vary, the proportions of "other" scrap and wastes remained stable or even decreased. This may indicate that the recyclables market in Vietnam is still in the "emerging" stage while that in the other countries is somewhat more mature.

The reasons for Vietnam's increase in "other" category recyclables include, first of all, the high demand for low-quality materials to produce low-quality goods. With a population of more than 86 million, many of whom have just escaped from poverty during the last 20 years, Vietnam still has a big market for low-priced, low-quality goods.[8] The second important factor is the lower costs of recycling. Mixed scrap is exported

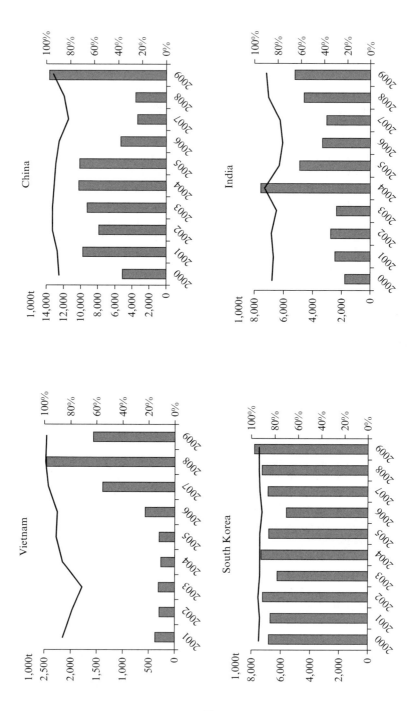

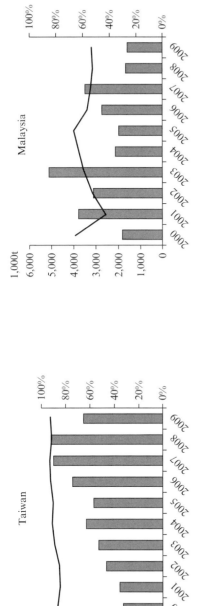

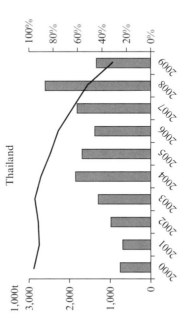

Source: World Trade Atlas.

Figure 4.4   Import volume of iron scrap and portion of "other" scrap in Asian countries

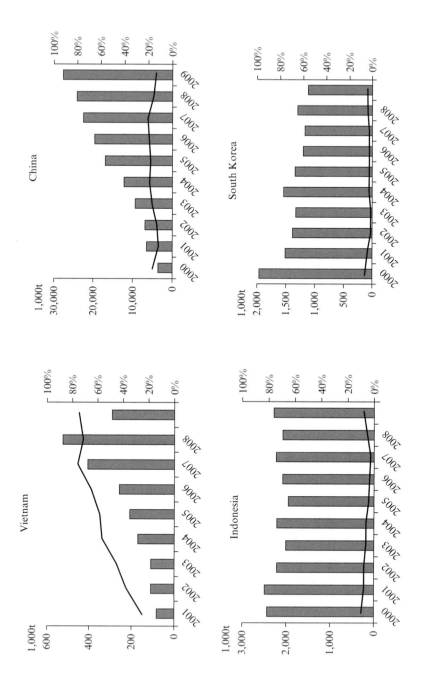

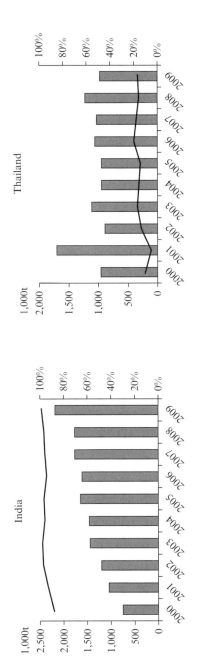

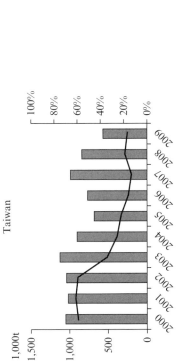

Source: World Trade Atlas.

Figure 4.5 Import volume of paper scrap and portion of "other" scrap in Asian countries

to Vietnam where segregation and dismantling can be conducted in a less costly manner by abundant low-wage manual laborers. Moreover, in Vietnam, where environmental regulations are still poorly enforced, residues (including hazardous wastes) after segregating or dismantling mixed wastes are easily disposed of with lower (or even no) environmental protection costs. In this sense, any increase in the import of "other" category recyclables must increase the risks of creating environmental problems.

4.2. REGULATIONS ON THE IMPORT OF RECYCLABLES TO VIETNAM

4.2.1 Regulations in the Environmental Protection Law

In Vietnam, the import of scrap and wastes as recyclable resources comes under the jurisdiction of the Environmental Protection Law and other legal orders under the Law. Some legal orders under the Commercial Law, which regulates external trade in general, also apply to the import of recyclables.[9] This is because the import of some recyclables is not only banned for the purpose of environmental protection, but is also restricted from the standpoint of the protection of domestic industries, consumer protection, or even national defense.

In the latest Environmental Protection Law, promulgated in 2005, the restrictions on recyclable imports are detailed in Articles 42 and 43. Article 42, which regulates the import of goods in general from the standpoint of environmental protection, is titled "Environmental protection in importation and transit of goods." The main message of this Article is simple and general: imports of goods that fail to comply with Vietnam's environmental standards are basically prohibited. Prohibited goods, if imported to Vietnam, must be re-exported, properly disposed of, or destroyed by their owners. In cases where the import of prohibited goods causes serious environmental problems, the owners are subject to administrative disposition or criminal prosecution. The Article also states that used machinery, equipment, and means of transport for the purpose of disassembling are also prohibited from import.

Article 43 covers the environmental regulations on the importation of recyclables (defined as "scrap" in these Articles). The full texts of Article 43 are presented in Appendix 4.1. "Scrap" is defined as "all products and materials that are discarded from a specific process of production or consumption but are collected as input materials for other productions." In other words, the wastes can be imported as scrap only if it is to be used as the input material for other production, not for the purposes of disposal in Vietnamese territory. The imported scrap, in principle, must be "segregated, cleansed and unmixed with materials, products and goods that are banned from import" (Clause 1).[10]

4.2.2 Government Decrees 80 and 81: Regulations under the Environmental Protection Law

In general, Vietnam's laws provide a basic direction for the regulations, which are usually followed by detailed regulatory guidance for implementation. The primary guidance for implementation of the 2005 Environmental Protection Law is in the form of Government Decrees No. 80 and No. 81 (80/2006/ND-CP and 81/2006/ND-CP; hereafter "Decree 80" and "Decree 81"), both issued in August 2006. In Decree 80, which covers a broad range of environmental issues, environmental regulations for external trade are referred to in Article 19 (see Appendix 4.2). However, Article 19 does not give any detailed instructions on environmental protection in import activities. The statements in the Article are as simple as those in Articles 42 and 43 of the Environmental Protection Law, with the exception of those on penalties for violators. Decree 80 seems to have been issued simply to pave the way for the Ministry of Natural Resources and Environment (MONRE) to issue ministerial-level legal orders such as Decision 12 (see Section 2.3).

Referred to in Clauses 2 and 3 of Article 19 are the regulations on the temporary imports of scrap with the purpose of re-export to other countries. Import of items banned for re-export is allowed only under the conditions that the packages are not opened and the imported volume is not reduced in Vietnamese territory (i.e., the imported items must be re-exported in their entirety).[11]

The detailed contents of the administrative dispositions are provided in Decree 81, which was issued the day after Decree 80. Illegal imports found to violate the Environmental Protection Law are subject to fines up to 70 million VND,[12] and the violators are forced to re-export goods that do not meet environmental standards and to remedy any pollution caused by the importation. With regard to the criminal prosecution of importers who violate the Environmental Protection Law, Article 185 of the Penal Code is applied (a maximum ten-year prison sentence).

4.2.3 Regulations on the Import of Recyclables under Commercial Law

Detailed guidance for implementing the Commercial Law is provided in Government Decree No. 12 (12/2006/ND-CP; hereafter "Decree 12"), which set forth import-export regulations including those concerning recyclables. The Decree specifies the commodities "banned from export and import," "licensed by the Ministry of Trade,"[13] and "subject to ministerial management and management regulations," and lists of these commodities are provided. Scrap in general, including iron and other non-ferrous metal scrap, waste plastic, and waste paper, are categorized as commodities to be "managed by the Ministry of Natural Resources and Environment" (MONRE) when imported. Second-hand machinery, motor vehicles, and electric appliances are listed in the category "banned from import."[14] Import of the items banned for re-export is allowed with permission from the Ministry of Industry and Trade (MOIT). The Decree also provides detailed stipulations on trade activities, such as licenses for traders and treatment of violators of the Law.

It was intended that Decree 12 would be followed by more concrete regulations from the responsible Ministries on those items categorized as "subject to ministerial management and management regulations." Some Ministries have in fact issued lists of the items banned from import (or approved for import), with HS codes. For example, the list of the approved scrap for import was issued as the Natural Resources and Environment Minister's Decision No. 12 (12/2006/QD-BTNMT, hereafter "Decision 12"). The Ministry of Post and Telecommunication[15] issued a list of used telecommunications equipment (TVs, personal computers, monitors, etc.) banned from import (20/2006/QD-BBCVT), and the Ministry of Transport listed the motor vehicles and motor vehicle parts banned from import (19/2006/QD-BGTVT).[16]

4.3   CASES OF ILLEGAL IMPORTS AND PROBLEMS IN ENFORCING THE REGULATIONS

4.3.1   Increases in Illegal Imports Causing Environmental Pollution

As can been seen, detailed regulations on the import of recyclables have been issued and penalties against acts of illegal importation have been more clearly defined since 2005. However, illegal imports continue. Or rather, the number of seizures of illegal imports has increased. However, as it is difficult to analyze the trends in environmental pollution caused by illegal imports based solely on information provided by MONRE and other related authorities, in this section we review several cases reported in local newspapers.

According to a report released by the "Environmental Police" (details given below) in November 2009, of the 2,575 cases of damage to the environment in the 22-month period of January 2008 to November 2009, more than 200 were related to illegal imports. The amounts of illegally imported goods that had required ship-backs or disposal by the importers were 325 tons of wastes, 10,000 tons of iron scrap, 3,150 tons of waste plastic, and 6,196 tons of lead-acid batteries (*Thoi bao Kinh te Viet Nam*, dated November 6, 2009).

Cases of large-scale illegal imports causing environmental pollution—in other words, explicit attempts to dump non-recyclable wastes (or hazardous wastes) in Vietnam—have been widely reported in the media since 2007. For example, in September 2007, 1,400 tons of imported iron scrap mixed with sludge and waste substances were found at Saigon Port in Ho Chi Minh City (*Lao Dong*, dated December 15, 2007). In January 2008, MONRE in cooperation with the Ministry of Finance, the Ministry of Public Security, the Ministry of Science and Technology, and the Vietnam Steel Association undertook investigations of imported iron scrap at Saigon Port and Hai Phong Port (Northern Vietnam), and seized 104 containers at Saigon Port and 157 containers at Hai Phong Port that were in violation of environmental regulations (*Thoi bao Kinh te Viet Nam*, dated January 14, 2008). In April 2008, 2,000 tons of imported iron scrap mixed with waste plastic and waste oil were found at Hai Phong Port (*Thoi bao Kinh te Viet Nam*, dated April 3, 2008). In 2009, there were two cases of seizure of large amounts of illegally imported lead-acid batteries (257 tons and 63,000 tons) in Quan Ninh Province, and a similar seizure of 44 tons in Hai Phong City (*Thoi bao Kinh te Viet Nam*, dated November 6, 2009).

Ship-backs did not occur in all cases of illegally imported scrap. In August 2008, 434 tons of iron scrap in 18 containers contaminated with arsenic and mercury were found at Tien Sa Port in Da Nang City (Central Vietnam). The importers claimed they were shipped from Italy, which the Italian embassy in Hanoi denied. Ultimately, these hazardous substances were incinerated in Vietnam at the importer's expense (*Sai Gon Giai phong*, August 10, 2008). In September 2008, at Tien Sa Port, the port authorities discovered ten containers of iron scrap abandoned for one year. These containers had been abandoned since the importer found that the scrap contained substances that broke the contract terms and refused to accept them. Local authorities in Da Nang City finally accepted the scrap for disposal (*VTC News*, dated September 18, 2008).[17]

4.3.2 Weak Capacity for Enforcing the Regulations

There are two possible interpretations for the increases in seizures. The growing number of seizures might be a result of increased attempts to import waste illegally; indeed, Vietnam is still regarded as a "pollution haven" where the risks of seizure and penalties are regarded as low. Or, it might be an indication of Vietnam's improved capacity to enforce the regulations. This study does not intend to analyze which interpretation is more plausible, but rather examines some of the problems behind the increasing trend in improper import practices.

The first problem observed is that the environmental regulations in Vietnam are sometimes unrealistically strict for domestic entities to abide by. Both the enforcement capacity of government agencies (especially of local government) and the financial capacity and awareness of many importers to obey the regulations are insufficient. It is true that, at the time the Environmental Protection Law was issued, Vietnam wanted to introduce new regulations because it already recognized that various environmental problems were unmanageable with existing laws. However, a more compelling underlying reason was the requirement to meet "international standards" in its laws and regulations. The environmental regulations have been developed under pressure from external forces. The Environmental Protection Law and Commercial Law were both enacted in 2005 along with several key economic laws, such as the Investment Law, Enterprise Law, and Intellectual Property Law. The development of these new laws was a condition to Vietnam's accession to the WTO. During the development of these laws, the Vietnamese government consulted with many overseas agencies and personnel to ensure the laws were acceptable to foreign investors and trade partners. As a result, these laws contain some new and foreign concepts, which require detailed institutional arrangements for their enforcement.

The second problem is that environmental standards pertaining to imported recyclable wastes are vague, making investigations of cases difficult. For example, the Environmental Protection Law, in principle, permits the import of scrap not "containing hazardous wastes and impurities" (Article 43.1.) and Decision 12 provides lists of the permitted recyclable scrap. However, no legal orders have specified concrete threshold levels for the hazardous wastes and impurities and, as such, the levels of "cleanness" or "dirtiness" of the imported scrap are judged subjectively at the investigation sites by the responsible authorities. Similarly, importers of scrap must have, according to the Environmental Protection Law, "adequate capacity of treating impurities mixed with scrap materials" (Article 43.2), but no guidelines have yet been provided to define the "adequateness" of treating impurities.

A further problem lies in the monitoring mechanism. Poor infrastructure is among the primary causes. In the above reported cases, non-recyclable and hazardous wastes mixed with recyclables were shipped to Vietnam in containers. It is more difficult to detect illegal shipments when the scrap is shipped in containers than when it is shipped as open bulk cargo. Cargo tends to be shipped in containers to Vietnam largely because of the lack of (or poor development of) port facilities. At present, the country has only a few ports, both public and private, that have the special facilities needed to load scrap from bulk ships. Consequently, many materials are shipped in containers to Vietnam. Importation in containers raises shipping costs in general. However, illegal shipping of banned wastes and scrap may enable the importers to generate profits that compensate for such costs. Moreover, importation in containers makes environmental inspections more difficult, which further increases the risks for environmental pollution.

4.3.3   Impacts and Limitations of the Environmental Police

Although enforcement of the environmental regulations remains plagued by a number of weaknesses, at least the capacity for seizure of illegal shipments has improved. The establishment of the Environmental Police Department in 2006 by the Ministry of Public Security has contributed to this improvement.[18] When a suspicious illegal act is detected, including the illegal import of banned substances at international ports, the local Department of Natural Resources and Environment (DONRE), the Customs Office, and the Environmental Police jointly investigate the case. Before the Environmental Police were established, the local DONREs were responsible for the investigations. However, the effects of the investigations were limited since DONREs do not possess the legal right to raid suspicious sites: in Vietnam, this activity remains firmly monopolized by the police authorities. The establishment of the Environmental Police has enabled investigations to be conducted more effectively and the regulations have become more enforceable.

The organizational capacity of the Environmental Police is still under development, however. Although the number of seizures has indeed increased recently, the Environmental Police are far from being capable of enforcing the regulations nationwide. The number of staff is still small and their activities are limited to the major international ports only. The expertise of the Environmental Police is also limited. As a unit, the Environmental Police have been separated from the Economic Police, and most of the staff were transferred from the Economic Police without being trained in environmental issues.[19] Coordination among the related actors involved is also cause for concern. Investigations have become more time consuming and costly with more actors involved than previously, with the local DONRE, the Customs Office, and the Environmental Police all participating in the inspections.

4.4 CONCLUSION

This chapter has explored the "dilemma" of the increasing inflow of recyclables (which constitute precious resources for industrial development) and the worsening of environmental pollution in Vietnam today. Import of recyclables to Vietnam has rapidly increased not only in terms of volume but also in the rate of mixed or low-quality scrap (categorized as "other" scrap), both of which could imply that improper imports may only increase into the future. Indeed, the number of the seizures of illegal imports increased during the latter half of the 2000s.

This chapter points out difficulties in establishing regulations that are suitable to protect the environment and also are enforceable with the existing institutional capacities. Vietnam introduced international standards in the processes of establishing new environmental regulations but had insufficient capacity in domestic entities to enforce the necessary regulations. Vietnam's integration into the global economy required the development of strict legal frameworks for such regulation, even though governmental agencies and the private sector have not yet been able to concretely implement the strict regulations. The current infrastructure is also insufficient to realize the high ideals of the environmental regulations.

Another concern—one that this chapter has not focused on—is that the environmental pollution caused by recyclable imports relates to improper recycling practices. Even legally imported scrap can cause environmental problems if it is improperly treated during recycling and final disposal of the residues. The speed of increase in the inflow of recyclables seems to have exceeded the speed of diffusion of costly modern technologies or proper management skills among the recyclers. It is becoming more difficult to manage the resulting situation because the demand for the imported scrap has increased from small-scale recyclers who are in many cases outside the government's control.[20] A similar structural problem, that is, strict regulations and insufficient mechanisms to implement them, can also be seen in the development of appropriate domestic recycling businesses.

A more fundamental problem behind the present situation in Vietnam concerns the decision-making processes. Regulations and restrictive orders are in general drafted by the policy makers and "specialists," without due consideration given to feasibility and the present weak implementation capacities. As a result, the current environmental regulations have come to contain some unrealistically strict provisions—these then lead to delays in the institutional development required for their implementation.[21] Nurturing the associations of recycling business entities and including these associations in the decision-making process would be a significant step toward lessening the gap between regulation and reality.

NOTES

1. The volume of waste plastic imported to Asian countries increased 100-fold, while imported metal scrap increased 10-fold from 1990 to 2008 (Michida 2011: 9).
2. According to Vietnam's Environmental Protection Law, "wastes" (*chat thai*) are defined as "materials that take a solid, liquid, gaseous, or other forms, are discharged from production, service, daily life or other activities" (Article 3.10). On the other hand, "recyclables" (*phe lieu*) refer to "all products and materials that are discarded from a specific process of production or consumption but are collected as input materials for other productions" (Article 10.13). In other words, recyclables are regarded as the resources or materials separated from the waste, to be used for production, and therefore are allowed for importation although import of wastes is, in general, prohibited. One more important definitional issue concerns second-hand goods. In Vietnam, second-hand goods such as used cars and electric appliances are not included in the category of recyclables. In general, the import of second-hand goods as recyclable materials is prohibited, and import for reuse is restricted to certain items such as used cars less than five years from production, and laptop computers.
3. http://www.gtis.com/twa.htm
4. Figures 4.2 and 4.3 present data for the exporting countries on export volumes to Vietnam. The reason why the import data are not used, as in the other tables and figures, is a problem of data limitation. The World Trade Atlas does not provide the complete time series data on imports to Vietnam by country.
5. Figures 4.2 and 4.3 show that the export volumes from major exporting countries to Vietnam did not decrease in 2009, which contradicts the data in Figure 4.1. The possible reasons for this discrepancy include the under-estimation of import volume in Vietnam's statistics, over-estimation of export data especially in the developed countries, and increase in re-export to other countries via Vietnam. Further studies, if more precise trade data are released, will be required to solve this problem. However, this contradiction does not deny the observation made here that the recyclable resources for Vietnam have been diversified.
6. Vietnam also signed a bilateral trade agreement (economic partnership agreement) with Japan in December 2008.
7   Total trade value (import plus export) in Vietnam increased more than five-fold from 30 billion USD in 2000 to 157 billion USD in 2010 (GSO 2011).
8. The poverty rate in Vietnam reduced from 58 percent in 1993 to 13 percent in 2008 (GSO 1994; 2009).
9. The Commercial Law does not include any articles that directly refer to the import of recyclables and wastes. Therefore, this chapter does not mention the contents of the Commercial Law, when reviewing, in Section 4.2.3, the legal orders under the Law.
10. One critical argument in the Law statements, when issued in 2005, was over the eligibility of the importers. There was no argument over the only capable entities that are allowed to import scrap being those who are equipped with proper facilities and technologies. However, Clause 2 of Article 43 stipulates that "organizations and individuals directly engaged in the use of scrap materials as input materials to their production and recycling processes shall be eligible to import scrap materials." This can be interpreted as implying that the trading enterprises, who do not usually directly produce anything from the imported scrap, would have been crowded out of the scrap import market. If so, the small-scale recyclers, who rely on trading enterprises for their material imports, would also have been seriously affected. This confusion was not resolved until the

Ministry of Natural Resources and Environment (MONRE) and the Ministry of Industry and Trade (MOIT) jointly confirmed, by issuing a Circular (02/2007/TTLT-BCT-BTNMT) in August 2007, that the "distributors" to the recyclers and those commissioned to import by the recyclers were also eligible to import scrap.

11. In September 2008, the MOIT issued an official letter (7893/BCT-XNK) ordering traders who import scrap with the purpose of re-export to comply with the two legal orders (12/2006/TT-BTNMT and 23/2006/QD-BTNMT). These orders issued in 2006 specify the ban on hazardous waste imports. This action by the MOIT implies that the regulations had not been well respected and that a considerable amount of temporarily imported goods categorized as hazardous wastes had entered into Vietnamese territory.

12. The fines for violation of the Environmental Protection Law were raised to, at most, 500 million VND by Government Decree No. 117 (117/2009/ND-CP, issued December 31, 2009) which took effect in March 2010.

13. In July 2007, the Ministry of Trade was merged with the Ministry of Industry to be reorganized into the Ministry of Industry and Trade (MOIT).

14. Second-hand motor vehicles less than five years from production are not subject to import ban. Among second-hand electric appliances, laptop computers were no longer banned from import in May 2007 following issuance of the Minister to Post and Telecommunication's Decision No. 11 (11/2007/QD-BBCVT).

15. The Ministry became the Ministry of Information and Communication in 2007.

16. As far as the author could determine from an examination of the legal documents, lists of the electric appliances banned from import, except for telecommunications equipment, have yet to be released (MOIT is responsible for the issuance). This may be the reason why imported second-hand electrical appliances such as refrigerators, washing machines, and audio equipment can often be observed in the market, although these are listed as items "banned from import" in Decree 12.

17. Illegal imports of non-recyclable, hazardous wastes have occurred not only in the importing of recyclable wastes. A state-owned steel enterprise in Hai Phong, which had won a tender for installing equipment to build a power plant in Nam Dinh Province, was subject to an administrative disposition because of its violation of the Environmental Protection Law. The company was alleged to have imported 40-year-old used machinery from South Korea, including electric transformers containing 4,000 liters of waste lubricating oil contaminated with PCBs (*Thoi bao Kinh te Viet Nam*, dated November 6, 2009).

18. The establishment of the Environmental Police Department was promulgated in Decision 1899 (1899/2006/QD-NCA), dated November 29, 2006.

19. In February 2009, when the author interviewed an officer of the Environmental Police, he revealed that the number of the staff at the headquarters in Hanoi was only about 60 and about 30 in the Ho Chi Minh office. Only three staff members in the Ho Chi Minh City Environmental Police have expertise in environmental issues.

20. In the Northern and Central parts of Vietnam, many recycling activities take place in "recycling villages" where rural villagers who moved from farming into the recycling business are forming clusters of recycling businesses (DiGregorio 1994; Dang Kim Chi ed. 2004; Sakata 2011). In these villages, manufacturing activities that use scrap and waste as materials take place on a village-wide (or commune-wide) scale. Most of the recyclers in these villages are family-run business entities who in many cases do not conduct business registration.

21. One example is the introduction of the extended producers' responsibility (EPR) principle for the collection of wastes. The introduction of the EPR principle is stipulated in the 2005 Environmental Protection Law, and after five years of preparation, the draft legal order (Prime Minister's Decision) was finally made public in early 2010. However, the draft prompted strong protests from the business community (especially foreign investors) because it contained unrealistic targets, such as a 75 per cent recycling rate for used and imported electric appliances. The draft was amended, to reflect (at least partly) public comments, and was submitted to the Prime Minister's Office in October 2010. The Decision has yet to be released at the time of writing this chapter.

REFERENCES

Dang Kim Chi (ed) (2004), *Lang Nghe Viet Nam va Moi Truong* (Craft Villages in Vietnam and Environment), Ha Noi: Nha Xuat Ban Khoa Hoc va Ky Thuat (Science and Technology Publishing House) (in Vietnamese).

DiGregorio, Michael (1994), *Urban Harvest: Recycling as a Peasant Industry in Northern Vietnam*, Honolulu: East-West Center.

GSO (General Statistical Office) (1994), *Vietnam Living Standards Survey 1992–1993*, Hanoi: Statistical Publishing House.

GSO (2009), *Result of Viet Nam Household Living Standards Survey 2008*, Hanoi: Statistical Publishing House.

GSO (2011), *Statistical Yearbook of Vietnam 2010*, Hanoi: Statistical Publishing House.

Imai, Kenichi (2009), 'Seisaku Katei to Sangyo Hatten: Tekkougyou no Kesu' (Policy Making Processes and Industrial Development: A Case of Iron Steel Industry)', in Tomohiro Sakaki (ed.), *Tenkannki no Chugoku* (China in Transition), Chiba: Institute of Developing Economies (in Japanese).

Kojima, Michikazu (2005), 'Tonan-Ajia Shokoku ni Okeru Junkan Shigen no Ekkyo Ido' (Trans Boundary Movements of Recyclable Resources in Southeast Asia)', in Michikazu Kojima (ed.), *Ajia ni Okeru Junkan Shigen Boueki* (Trade of Recyclable Resources in Asia), Chiba: Institute of Developing Economies (in Japanese).

Kojima, Michikazu and Shozo Sakata (2008), 'Betonamu no Haikibutsu/Risaikuru Kisei no Saishin Joho' (The Latest Trends of Regulations on Waste Management and Recycling in Vietnam), *World Eco Scope 2009*, Tokyo: Daiichi Hoki (in Japanese).

Kojima, Michikazu and Aya Yoshida (2007), 'Betonamu ni Okeru Sangyo Haikibutsu Risaikuru Seisaku' (Policies on Industrial Waste Management and Recycling in Vietnam), in Michikazu Kojima (ed.), *Heisei 18 Nen-do Ajia Kakkoku ni Okeru sangyo Haikibutsu Risaikuru Seisaku Joho Teikyo Jigyo houkokusho* (FY 2006 Project Report on Information Provision on Policies on Industrial Waste Management and Recycling in Asian Countries), Chiba: Institute of Developing Economies (in Japanese).

Michida, Etsuyo (2011), 'International Trade of Recyclables in Asia: Is Cross-border Recycling Sustainable?', in Michikazu Kojima and Etsuyo Michida, *Economic Integration and Recycling in Asia: An Interim Report*, Chiba: Institute of Developing Economies.

Pham Thai Hung, Bui Anh Tuan and Nguyen The Chinh (2008), *The Impact of Trade Liberalization on Industrial Pollution: Empirical Evidence from Vietnam*, EEPSEA Research Report, Singapore: Economy and Environment Program for Southeast Asia.

Sakata, Shozo (2011), 'Clusters of modern and local industries in Vietnam', in Ikuo Kuroiwa (ed.), *Spatial Statistics and Industrial Location in CLMV*, Chiba: Institute of Developing Economies.

APPENDIX 4.1:   ARTICLE 43 OF THE 2005 ENVIRONMENTAL PROTECTION LAW

Article 43:   Environmental Protection in Importation of Scrap Materials

1. Scrap materials to be imported must comply with the following requirements for environmental protection:
   (a) Having been segregated, cleansed and unmixed with materials, products and goods that are banned from the import in accordance with the provisions of the law of Viet Nam or international treaties to which the Socialist Republic of Viet Nam is a Contracting Party;
   (b) Without containing hazardous wastes and impurities, except non-hazardous impurities mixed during loading, unloading and transport operations;
   (c) Falling under the list of scrap material categories that are permitted for import established by the Ministry of Natural Resources and the Environment.

2. Organizations and individuals directly engaged in the use of scrap materials as input materials to their production and recycling processes, shall be eligible to import scrap materials if the following terms and conditions are satisfied:
   (a) Having storehouses and yards separate for the storage of scrap materials to ensure conditions for environmental protection;
   (b) Having adequate capacity of treating impurities mixed with scrap materials;
   (c) Having technologies adopted and equipment provided for recycling and reusing scrap materials to meet the environmental standards.

3. Organizations and individuals engaged in the import of scrap materials shall have the responsibility to:
   (a) Comply with the regulations by the law on environmental protection and other provisions of the relevant law;
   (b) Notify, in writing, the State management agencies of environmental protection at provincial level where their production units, storehouses or yards of imported scrap materials are located, within at least five (5) days before the date of loading or unloading scrap materials, of categories, quantities and weights of imported scrap materials, border gates, transport routes, storehouses and yards for storing scrap materials and places where scrap materials are fed to production;

(c) Implement the treatment of impurities mixed with scrap materials; and handing over or sale of impurities are prohibited.

4. Provincial level People's Committees shall have the responsibility to:
    (a) Inspect, detect, prevent and deal with acts of violation against the law relating to imported scrap materials;
    (b) Annually report to the Ministry of Natural Resources and the Environment on the situation of the import and use of scrap materials and imported scrap materials related environmental issues in their localities.

5. The import of scrap materials is a conditional form of business. The Ministry of Trade shall, in collaboration with the Ministry of Natural Resources and the Environment, be primarily responsible for imposing criteria and conditions on the business of organizations and individuals engaged in the import of scrap materials.

APPENDIX 4.2: ARTICLE 19 OF THE DECREE 80 (80/2006/ND-CP, AUGUST 9, 2006)

Article 19: Environmental Protection in the Import, Temporary Import, Border Gate-to-border Gate Transport and Transit of Scrap

1. If organizations and individuals importing scrap fail to comply with the provisions of Clause 1 and Clause 2, Article 43 of the Law on Environmental Protection, they shall, depending on the nature and severity of their violations, be administratively handled or examined for penal liability. If causing damage, they shall have to compensate therefore according to the provisions of law.

2. Temporary import and border gate-to-border gate transport of scrap shall have to strictly meet the following requirements:
    (a) Scrap must not be unpacked, used and dispersed in the course of transportation and storage in Vietnam;
    (b) Scrap must not be altered in nature and weight;
    (c) Scrap must all be re-exported or transported out of Vietnam.

3. Transit of scrap through the Vietnamese territory shall have to meet similar environmental protection requirements prescribed in Article 42 of the Law on Environmental Protection for transit of goods.

5. South Korea's approach to transboundary waste management: experiences and lessons learned

Sungwoo Chung

INTRODUCTION

Countries take different positions in their policies on the movement of transboundary waste depending on their respective domestic waste–related situation.[1] Asian countries such as China, Indonesia, and Vietnam actively take legal action to prohibit the import of hazardous waste for fear of pollution caused by improper treatment. In particular, for second-hand goods such as waste electrical and electronic equipment (waste EE equipment), a variety of policy positions are evident, ranging from a complete ban of imports to selective acceptance with consideration given to year of production. Overall then, awareness of the need for effective use of this transboundary waste and its hazard management is spreading.

Since around 2004, the reduction of trade barriers for transboundary waste has been newly discussed from the perspective of promoting appropriate international resource circulation (Kojima 2011, p. 27). However, it is still a challenging question as to how we balance environmental protection and economic efficiency in transboundary waste management given that the boundary between hazardous waste and secondary goods can be unclear.

To manage the above situation, South Korea has until recently relied on the Act on Waste Management (AWM) and the Act on the Transboundary Movement of Waste and its Treatment (ATW), which have been the vehicles for domestic implementation of the United Nations' Basel Convention. However, the mismatch in the above regulations has resulted in soil pollution in Kangwon Province, which signifies that they failed to appropriately manage the increasing amounts of transboundary waste entering South Korea since the late 1990s (see Section 5.1). As a new countermeasure, South Korea introduced the Export/Import Declaration

System (EIDS) in 2008 in a revision to the AWM. This system is unique in that it attempts to grasp the general flow of non-hazardous transboundary waste and applies to all forms of transboundary waste. Thus, how effectively this system functions requires careful review in order to extract policy implications; however, only limited research has been carried out on this issue to date.

In this chapter, the author reviews the background on the legal framework and actual implementation of the current regulatory measures on transboundary waste in South Korea. This review will highlight the policy challenges and remedial action necessary for more sustainable transboundary waste management. Section 5.1 provides a general introduction to the South Korean regulatory system and gives definitions of waste and an outline of the legal framework for transboundary waste in South Korea. Section 5.2 explains actual practice as South Korea tries for more sustainable management of transboundary waste. In Section 5.3, policy suggestions based on South Korea's experiences to date are provided in two main areas: the linkage between the regulations of waste and improvement measures for more sustainable waste management.

5.1   SOUTH KOREA'S REGULATORY SYSTEM FOR TRANSBOUNDARY WASTE

5.1.1   Definitions of Waste and Second-hand Goods

According to the AWM, which governs the proper treatment of domestic solid waste, waste is defined as "material that is unnecessary for human life and business activities such as garbage, combustible ash, sludge, waste oil, waste acid, waste alkali, carcasses, etc." (Article 2, paragraph 1). In reality, the decision on whether a certain material legally constitutes waste is ultimately left to the courts.[2] However, in this chapter waste-related issues are discussed on the basis of the administrative interpretation of the Ministry of Environment (MOE), which is responsible for the enforcement of the AWM.

On the basis of the provisions above, it can be said that South Korea's regulation places a high value on the viewpoint of the discharger in identifying waste. In other words, what is produced as unnecessary goods by the discharger is preferentially treated as waste, irrespective of its usefulness to a third party.[3] If defective products (returned goods, etc.) are generated by the same facility where it is utilized as raw material, it is not regarded as waste from a legal perspective. However, in cases where defective products are used as raw material by a third company, it should be handled as waste (MOE 2004, p. 7).

In regard to transboundary waste, the ATW stipulates that waste is defined as a material that is listed in the Annex of the Basel Convention or that requires import and export regulation through a bilateral, multi-lateral, or regional convention (Article 2, paragraph 1). The specific waste list has been announced as a presidential decree (refer to Section 5.1.2). Recycling is defined as "activities of reusing/using after regeneration or making waste reusable/usable after regeneration" (ATW, Article 2, paragraph 7). Recycling activity is activity involving waste from the legal perspective, irrespective of the degree of economic value.[4] As for reuse, South Korean courts have delivered the judgment that reuse is one type of recycling. In South Korea, second-hand goods are basically treated as one type of waste and reuse is one example of recycling from the legal perspective.

5.1.2   Regulations to Manage Transboundary Waste and Respective Target Waste

As mentioned above, South Korea has enacted two main acts to effectively cope with the transboundary movement of waste, the AWM and ATW. In September 2008, the EIDS was newly introduced as an additional measure through revision of the AWM to prevent the illegal export of waste and secure the proper treatment of imported waste. This measure takes a unique and challenging approach to the issue in that the system covers both hazardous and non-hazardous waste comprehensively. With the adoption of this new system, transboundary waste is categorized into three groups: waste requiring approval on the basis of the ATW, waste requiring declaration on the basis of the AWM, and waste that does not require approval or declaration for transboundary movement.

The specific waste items regulated by the ATW are announced by the MOE through public notice. The most recent MOE notice (No. 2007-188) lists 86 items as waste requiring import and export approval (WRA). In comparison to the Basel Convention and OECD rules, target waste includes 61 items from Annex I of the Basel Convention, 2 items from Annex II of the Convention, and 23 items from the list of green waste in the OECD rules. Similarly to the ATW, under the AWM, 25 items are listed in the MOE notice as waste requiring import and export declaration (WRD) (Table 5.1). If waste falls under both the ATW and AWM, it is treated as WRA to be regulated by the ATW. The 25 items requiring declaration are treated as general industrial waste in the AWM. It is not permitted to export such waste without proper treatment after import.

Table 5.1   Listed waste requiring import and export declaration (WRD)

1 Waste synthetic polymers	13 Combustion residues
2 Sludge	14 Waste stone
3 Mineral ash	15 Waste tires
4 Dust	16 Waste cooking oil
5 Waste refractories and ceramic fragments	17 Animal and plant residues
6 Incineration ash (bottom ash/fly ash)	18 Waste EE equipment
7 Residues from stabilization and solidification	19 Chaff and bran
	20 Waste wood
8 Waste catalysts	21 Waste sand
9 Waste adsorbents and waste absorbents	22 Waste foundry sand
10 Waste paint and lacquer	23 Waste fiber
11 Waste plant and animal oil	24 Waste metal
12 Waste plaster and waste lime	25 Waste glass

Source: MOE notice (No. 2010-57).

Transboundary waste which does not require approval or declaration is categorized into six groups (MOE 2010, pp. 3–5). The first group concerns material for regeneration or recycling products; for example, wood pellet made from sawdust is treated as a recycled product, which is traded as common goods. The second group is waste traded as products internationally, such as silica fume from concrete hearths and borazon from melting coal slag. The third group is waste paper and metal scrap, which does not require any report to the MOE under the AWM; however, if an extraneous substance is added or waste oil is contained, the material falls outside this categorization. The fourth group concerns items recognized as environmentally friendly products by the International Standardization Organization (ISO). The fifth group is recycled products recognized by South Korean environmental standardization including South Korean industrial standards. The sixth group is what we call second-hand goods. Although second-hand goods are generally exported and imported as common goods, distinguishing them from WRD is not a simple issue. To address this problematic situation, the MOE suggests several standardizations to enhance the effectiveness of the regulations (Table 5.2). They are mainly composed of substantial standardizations, not specifying the production time and so on.[5] This can also be understood as South Korea's approach to the circulation of second-hand goods.

Table 5.2.   Standards for second-hand and waste EE equipment

Second-hand EE equipment	Waste EE equipment
Fully functioning	Imported for the purpose of recycling
Attachment of contracts regarding	(not reuse) or disposal
the sale and transportation for the	Difficulty with normal functioning
purpose of reuse	from parts failures or physical
Attachment of evaluation documents	damage
or test result issued by an official	Use of parts prohibited by Korea's
institution in the exporting country	domestic law
	EE equipment only for parts use

Source: MOE (2010).

5.2   ACTUAL PRACTICE OF REGULATION FOR TRANSBOUNDARY WASTE

5.2.1   Import and Export of Transboundary Waste

As a general trend in the trade of hazardous waste under the ATW in South Korea, the import volume of hazardous waste is much greater than that of exports (Tables 5.3 and 5.4). Additionally, it is noteworthy that the total volume of hazardous waste imports in 2009 was more than three times that in 2005. The current increase in the volume of traded waste (particularly imports) underscores that strengthening the regulations governing the waste trade is a necessity in South Korea. Specifically, in the mid-2000s large amounts of PCB-containing waste transformers and waste PCB oil became the main types of export waste, bound for the Netherlands and France; in the late 2000s waste EE equipment and car debris became the main types exported under the ATW.

In terms of import volume, the primary imports in the late 2000s were sludge, waste batteries, and lead lumber. In particular, waste batteries have been continually imported, making up more than 95 percent of the total import volume in 2009 (Table 5.4). The leading exporters of these materials are reported to be the United States (waste batteries) and Japan (lead-acid battery waste). The remaining imported waste, excluding waste batteries, includes waste CRT (glass-tube) televisions, the number of which is expected to decline considering that there is only one company manufacturing them and no longer any demand for CRT televisions in South Korea (Min Dal-gi 2009, p. 101).

Table 5.3   Export of hazardous waste under the ATW, 2006–09 (unit: tons)

	2005	2006	2007	2008	2009
Waste transformers containing PCB	101	407	388	56	–
Waste oil containing PCB	373	151	–	–	–
Waste plastic	13	37	–	–	–
PVC cable	278	–	–	–	–
Waste alkali	–	–	14	–	–
Waste EE equipment	–	–	–	220	552
Car debris	–	–	–	–	34
Total	765	595	402	276	586

Source: MOE (2011).

Table 5.4   Import of hazardous waste under the ATW, 2005–09 (unit: tons)

	2005	2006	2007	2008	2009
Sludge	823	2,727	769	2,038	180
Waste batteries	44,806	57,830	122,362	134,622	142,568
Ni-Cd	89	186	295	173	1,012
Lead scrap	2,187	119	–	330	–
Lead lumber	260	102	1,035	2,127	3,213
CRT glass tubes	–	152	–	–	1,636
Waste EE equipment	–	–	–	151	27
Slag	–	–	–	17	17
PVC cable	–	–	–	–	35
Total	48,165	61,116	124,461	139,458	148,688

Source: MOE (2011).

From late 2008 to early 2009, an average of 1 export transaction and 3.5 import transactions per month have been reported (MOE 2009, pp. 13–14). In 2009, export of non-hazardous waste was less than one-tenth that of import (Table 5.5). While waste synthetic polymers account for most exports, incineration ash accounts for most imports of non-hazardous waste in South Korea. Specifically, waste EE equipment and waste synthetic polymers are primarily exported to China. As for imports, incineration ash from Japan and waste vehicle catalyst from the United States and Colombia are typical imported items.

Apart from transboundary waste that requires no declaration or

Table 5.5   Export and import of non-hazardous waste under the AWM (unit: tons)

	Export		Import	
	2008	2009	2008	2009
Waste synthetic polymers	4,754	17,137	1,453	50,888
Waste catalysts	167	5,974	1,897	3,827
Waste tires	–	5,843	3,621	14,901
Slag	4,110	8,769	–	–
Dust	11,800	640	–	–
Waste glass	77	5,125	–	–
Waste EE equipment	172	2,463	–	835
Animal and plant residues	–	3,630	–	–
Sludge	–	9,344	–	–
Waste metal	–	1,695	–	73
Waste fiber	–	855	–	44
Waste cooking oil	–	–	183	3,672
Incineration ash	–	–	393,475	792,448
Total	21,080	61,475	400,629	866,688

Source: MOE (2011).

Table 5.6   Volume of transboundary waste (2009) (unit: tons)

	Export	Import	Total
WRA (waste requiring export and import approval)	586	148,688	149,274 (13.8%)
WRD (waste requiring export and import declaration)	61,475	866,688	928,163 (86.2%)
Total	62,061	1,015,376	1,077,437 (100%)

Source: Compiled by author on the basis of MOE (2011).

approval, the volume of transboundary waste amounts to approximately 1.07 million tons (Table 5.6). About 86.7 percent of transboundary waste is managed by the EIDS, which signifies the increasing importance of the AWM. As confirmed in the previous section, import of incineration ash accounts for 73.5 percent of the total transboundary waste in South Korea.

5.2.2 Implementation of the EIDS and Distinguishing Target Waste

As mentioned in the Introduction, because waste imports have rapidly increased since the late 2000s, concerns about environmental pollution have intensified accordingly. In response, the South Korean government introduced the EIDS to prevent the export of illegal waste and promote the proper treatment of imported waste.

Exporters of waste requiring a declaration (i.e., WRD) must provide the following three documents to the local environmental agency (LEA): a copy of the order, including export price based on the freight on board; the transfer plan for the export waste; and results of the analysis of the export waste issued by an authorized testing laboratory. If changes must be made to the documentation, the documents must be resubmitted to confirm the alterations. If there is a change in the volume of export waste (more than a 30 percent volume increase for designated waste and more than a 50 percent volume increase for other types of waste), this change must also be reported to the LEA. In addition, importers of WRD must report the location of the treatment facility to the LEA in the following three documents: a copy of the order, including import price based on the cost, insurance, and freight; the transfer and treatment plans for the imported waste; and the analysis results of the imported waste. WRD shall not be re-exported without first being properly treated in South Korea. Imported WRD are transferred, stored, and treated in accordance with the standards and treatment methods for industrial waste in the AWM. The process of acceptance and take-over of the waste is maintained by the manifest system, known as the Allbaro system.

In order to increase the effectiveness of the EIDS, penalties and fines may result from violations of the required procedures. Persons who place imported WRD in a landfill and export WRD without prior treatment may be sentenced to at least three years in prison or fined up to 20 million won. In addition, persons who pollute the environment in the course of transferring and storing WRD can receive at least two years in prison or a fine of up to 10 million won.

In 2010, to solve the problematic situation of which regulation should be applied for each type of transboundary waste, a guideline was published by the MOE (Table 5.7). First, waste plastic corresponds basically to the definition of WRD, but in the case of it being mixed with hazardous waste or contaminated by pollution, it falls under the jurisdiction of the ATW.[6] Pure metals, ferroalloys, and nonferrous metals are exempt from declaration. Iron oxide in powder form must be declared. However, scrap containing hazardous materials (e.g., stibium and arsenic), cutting chips contaminated with cutting oil, and automobile parts are classed as WRA.

Table 5.7 Distinctions between different types of transboundary waste

	Waste export import declaration system	Waste export import approval	Note
Waste plastic	Normal	Mixed with hazardous waste/ contaminated with pollutants	Declaration or approval is unnecessary when used as a material such as pellet or flake
Waste metal	Iron oxide in powder form	Scrap containing hazardous material (stibium, etc.), cutting chips contaminated with cutting oil	Declaration or approval is unnecessary for pure metals, ferroalloys, and nonferrous metals
Waste EE equipment	Scrap (plastic, metal, etc.) generated in the manufacturing process	Printed circuits	Waste transformers (including second-hand goods) are not permissible for export or import
Waste catalyst	With transition metals or rare earth metals	If containing palladium or platinum, becomes WRA	–
Coal ash	General coal ash (fly ash/bottom ash) generated in a thermoelectric power plant	Fly ash containing hazardous material	–

Source: Compiled by author on the basis of MOE (2009) and MOE (2010).

Waste tires imported to be utilized as a material for regenerated tires are categorized as WRD. PE and PVC waste cable is categorized as WRD, but waste cable containing coal tar, PCB, or cadmium is covered by the ATW. As for waste EE equipment, waste printed circuits are classed as WRA; however, scrap (mainly plastic or metal) generated in the process of manufacturing is classed as WRD. Waste transformers are not permissible for export or import irrespective of PCB content. Silica fume used to strengthen fire bricks for concrete hearths does not require any declaration or approval. Waste batteries for automobiles, in principle, are WRA. Steel slag is considered to be a type of mineral ash, although approval is required if it contains any hazardous material. Waste catalysts containing transition metals or rare earth metals should be managed in accordance with the AWM, but it is classified as WRA if it contains palladium or platinum.

Because coal ash (fly ash/bottom ash) generated in a thermoelectric power plant is classified as combustion residue it should be declared (MOE 2008b, p. 96). Fly ash containing any hazardous material must be approved for export or import.

5.3 DISCUSSION

5.3.1 Link between Regulations and Strengthening the Pre-trade Stage

In the AWM, waste is chiefly classified as household waste and industrial waste on the basis of where it is generated. Industrial waste includes waste that is notably hazardous to human health, medical waste generated by medical institutions, and other types of dangerous material. In addition, designated waste is treated as hazardous waste, for which there are 11 categories according to the presidential decree.

Under the AWM, designated waste is regulated as waste that must be properly treated. South Korea's system specifies only that transboundary waste is industrial waste, which results in there being no clear relation between the ATW and regulations for transboundary waste. The broad relations between these two classifications are shown in Table 5.8. Waste EE equipment and chaff/bran should be treated as industrial waste under the EIDS, but the AWM does not specify them as waste to be managed. This inconsistency in regulations often results in the illegal export of designated waste and improper treatment of WRA.

Specifically, only 53 waste items out of 86 classified as WRA are managed as designated waste under the AWM. The remaining 33 items are just classified as general industrial waste, not designated waste. It

Table 5.8 Listed WRD and relations with the waste classification in the AWM

Waste requiring import and export declaration	Classification of waste in AWM			
	Household waste	Industrial waste	Construction waste	Designated waste
1. Waste synthetic polymers	Waste plastic	Waste synthetic polymers	Waste synthetic resins	Waste synthetic polymers
2. Sludge	–	Sludge	–	Sludge
3. Mineral ash	–	Mineral ash	–	Mineral ash
4. Dust	–	Dust	–	Dust
5. Waste refractories and ceramic fragments	Metal hyaline	Glass ceramics	–	Waste refractories and ceramic fragments
6. Incineration ash (bottom ash/fly ash)	–	Incineration ash	–	Incineration ash
7. Residues from stabilization and solidification	–	–	–	Residues from stabilization and solidification
8. Waste catalysts	–	Waste catalysts	–	Waste catalysts
9. Waste adsorbents and waste absorbents	–	Waste adsorbents and waste absorbents	–	Waste adsorbents and waste absorbents
10. Waste paint and lacquer	–	–	–	Waste paint and lacquer
11. Waste oil	–	–	–	Waste oil
12. Waste lime plaster	–	Waste lime plaster	–	–
13. Combustion residues	Waste coal briquettes	Combustion ash	–	–

				Residential construction waste
14. Waste stone	—	—	—	—
15. Waste tires	Rubber, leather	—	—	—
16. Waste cooking oil	—	Waste cooking oil	—	—
17. Animal and plant residues	Food waste	Animal and plant residues	—	—
18. Waste electrical and electronic equipment	—	—	—	—
19. Chaff and bran	Food waste	Animal and plant residues	—	—
20. Waste wood	Waste wood	Waste wood	Waste wood	—
21. Waste sand	Waste sand	—	Waste sand for construction	—
22. Waste foundry sand	—	—	Waste foundry sand	Waste foundry sand and sandblast
23. Waste fiber	Combustible fiber	Waste synthetic fabric	Waste fiber	—
24. Waste metal	Metal, hyaline	Waste metal	Waste metal	—
25. Waste glass	Metal, hyaline	Glass ceramics	Waste glass	—

Source: Min Dal-gi (2009).

is necessary for at least some of these 33 items to be included as WRA because they could cause pollution if treated improperly. One possible concrete measure would be to gradually expand the coverage of designated waste, taking into account the situation of relevant companies and the level of hazard, by classifying waste into three groups: red waste to be verified internationally (first stage), green waste for which technical guidelines are prepared (second stage), and the other waste (third stage) (Lee Seung-hi et al. 2008, p. 299). More importantly, finding a way to ensure consistency between the regulations of domestic-generated waste and transboundary waste is clearly a policy challenge.[7]

Despite the existence of the MOE guideline for determining which regulation should be applied to which aspect of waste trade, there are still considerable amounts of WRA that are imported or exported without going through the necessary procedures. From January 2004 through October 2006, various types of WRA have been exported without prior notice and consent from importing countries, and these included waste oil (55 million liters), waste organic solvents (20 thousand liters), and waste catalysts (1,320 tons) (*Hangyere* 2006). In South Korea, provisions at the pre-trade stage are not sufficient to prevent the illicit movement of transboundary waste. Determination of which regulation applies primarily rests with importers and exporters. If they consider imported or exported waste as not requiring approval as WRA, they proceed to trade it as common goods, which are not required to go through the procedures for WRA.

To prevent the export of WRA disguised as second-hand goods, measures that strengthen the determination stages, such as the "prior consultation" stage operated in Japan,[8] would be effective. The reporting requirements for frequently traded waste with a high potential for pollution need to be reviewed because it is inefficient to require reporting for all WRA. In addition, to increase the effectiveness of the regulations on waste import and export, regulations at the customs stage also need to be strengthened. The South Korea Customs Service issued a revised public notice (No. 2009-115) in 2009 to tighten regulations for the import and export of waste. However, the tightened regulations only applied to changes made to the lists of waste and the addition of waste requiring regulation. These steps do not include the HS codes that would enhance the effectiveness of the regulations (Kim Yong-won 2008, pp. 156–7). In fact, customs civil servants manage waste by matching the inventory list and number on the declaration form, which is neither an efficient nor an effective method for regulation. Neither clear standards for evaluating the hazardousness of WRA nor data from the dissolution test are specified, resulting in a problem of regulatory uncertainty.

5.3.2   Movement of Transboundary Waste and Environmental Pollution

In South Korea, coal ash generated from thermal power plants is usually sold at a positive price to cement manufacturing companies, which utilize it as raw material for producing cement. During the economic crisis of the late 1990s, the construction industry fell into a recession, which subsequently caused an oversupply of cement. To cope with this problem, the cement industry demanded that the MOE deregulate the use of industrial waste in kilns. In response, the MOE approved the comprehensive usage of industrial waste such as coal ash in kilns. The MOE thought that utilizing industrial waste as kiln fuel would be beneficial to increasing recycling activities as well as improving management of the cement industry. As usage of industrial waste in kilns increased, the recycling rate of industrial waste also increased accordingly, from 62.9 percent in 1997 to 83.1 percent in 2005 (MOE 2008a, pp. 534–6).

However, despite the growing importance of the recycling of industrial waste in kilns, the AWM did not provide sufficient installation standards, such as those pertaining to exhaust gases, for kilns in waste recycling and treatment facilities. This deficiency was the main reason that monitoring of the type and volume of waste disposed in kilns was not efficiently undertaken. Through these insufficient and irrelevant regulations, kilns have caused ill health among residents in Kangwon Province since the mid-2000s (Choi Byung-sung 2007, pp. 14–18; Korea Chemical Research and Development Institute 2007, pp. 23–8).

Under this regulatory regime, the import volume of coal ash from Japan to South Korea has drastically increased since 2003. This trend provoked increasing concerns about the hazards of coal ash, which were eventually discussed in South Korea's National Assembly. Reflecting the discussions held in the National Assembly, the EIDS was introduced in 2008 with the purpose of strengthening the management of imported hazardous waste. Consequently, the growing import of coal ash from Japan was a crucial factor in strengthening the regulation of waste trade in Korea.

According to the ATW, steel slag and coal ash generated in thermal power plants are not considered to be WRA. However, fly ash, which contains hazardous materials greater than the standards set forth in the Basel Convention, should be regulated as WRA. For fly ash, according to the OECD rules it can be traded between member countries without restrictions. As for coal ash, which includes various toxic substances, it was actually traded between South Korea and Japan as a commodity as mentioned above and prompted a national dispute over the environmental pollution caused by treating these coal ash imports.[9]

Next, let us consider the impact that coal ash imports from Japan

Table 5.9   Volume of coal ash imports from Japan and fly ash incineration in South Korea (unit: 1,000 tons)

		2002	2003	2004	2005	2006	2007	2008	2009	2010
Import volume of coal ash from Japan	Korea statistics	93	289	414	513	715	689	857	792	N/A
	Japan statistics	N/A	N/A	N/A	N/A	624	600	763	784	941
Incineration volume of fly ash in Korea		635	445	1,117	1,639	1,196	960	N/A	N/A	N/A

Source: BAI 2009; MOE 2008c; MOEJ (www.env.go.jp).

have had on South Korean landfill sites, in addition to the environmental hazards they create. Before the full-scale import of coal ash from Japan, most fly ash generated in South Korea was recycled in kilns, and bottom ash containing highly toxic material was disposed of in landfill sites. However, to make room for the more than 500,000 tons of coal ash imported from Japan, fly ash generated in South Korea was no longer recycled in kilns (Table 5.9). In other words, it was more profitable for South Korean cement manufacturing companies to treat the imported coal ash from Japan because about 50,000 won per ton is paid in disposal costs, which is higher than that paid by South Korean power plants to the cement companies. Thus, the import of coal ash from Japan has played a substantial role in depleting South Korean landfill sites, as well as making the recycling of coal ash in South Korea more problematic. This case shows that attention must be given to the impact of transboundary waste on domestic recycling and landfill sites. In the Basel Convention, transboundary movements of hazardous waste destined for landfill sites are explicitly banned. South Korea's situation needs to be viewed as an interesting and cautionary one: landfill sites can be decreased in an importing country by the promotion of waste trade even between advanced countries.

5.4   CONCLUDING REMARKS

In this chapter, the regulations of waste trade in South Korea and the policy challenges it faces are discussed. South Korea has established a two-tiered system to effectively manage the import and export of hazardous waste: the ATW and AWM, which serve to implement the Basel Convention. In particular, transboundary waste, excluding hazardous waste, is controlled by the relatively new EIDS which is under the jurisdiction of the AWM.

Despite these recent moves toward more effective control of the waste trade, there are still gaps between regulation and actual practice. There is as yet no uniformity in the regulations for handling hazardous waste classified in the ATW and AWM or for managing WRD. Effective management of hazardous waste imports at the customs stage is necessary because HS codes do not cover the 86 items of WRA. Moreover, there is a lack of regulation for WRA, which is not included as designated waste.

As an additional point, the fact that determining the applicable regulation for waste rests primarily with importers and exporters is also a policy challenge. Strengthening the pre-trade stage is necessary for more effective regulation, and the legal definition of second-hand goods should be clarified with consideration given to the proper treatment of transboundary waste.

Finally, as in the case of coal ash imported from Japan, more discussion is needed about what kind of policy response would be most effective in dealing with the shortage of landfill sites caused by the transboundary movement of waste. This landfill issue touches on a number of concerns in addition to that of the environmental pollution caused by imported and exported waste, which must be addressed. When seeking a solution to such a situation, the potential effects of waste trade on the status of recycling in the importing country become a very important factor to consider.

NOTES

1. Waste is generally categorized as hazardous waste and non-hazardous waste for purposes of the transboundary movement of waste. However, South Korea has put more policy emphasis on effective monitoring of waste flow rather than complete control of hazardous waste between countries. In this sense, the author has decided to accept the term of "transboundary waste" when analyzing South Korea's case.
2. Fertilizer manufacturer A stored liquid fertilizer in a factory with insufficient leak prevention facilities. During a rainy period, the fertilizer was discharged outside of the factory and polluted neighboring farm land. The court ruled in this case that the fertilizer was waste because it became unnecessary once discharged, irrespective of its storage for future use.
3. For example, in cases where material is generated and discarded in the manufacturing process and can be provided to a third party for recycling, there are legal obligations required, such as declaring the material recyclable waste, because it is otherwise considered to be waste by default (Kim Hong-gyun 2007, p. 257).
4. Japan regulates that it is recognized as waste only when the flow of an item and currency is consistent.
5. The Japanese government provides criteria that TV sets are recognized as second-hand goods only if they are less than 15 years from production.

6. When waste plastic is used as a material such as pellets or flakes, declaration or approval is unnecessary.
7. For example, when foreign-generated transboundary waste belongs to the list of ATW but is not designated as hazardous waste, proper treatment is not guaranteed. In addition, a similar problem would occur if clear standardization on hazardous components is not harmonized.
8. However, the system of "prior consultation" in Japan uses only documents filed in relation to import and export transactions for decision making. This system does not prove that importers and exporters comply with the relevant laws in terms of trade (Tsuruta and Yoshida 2009, p. 61).
9. In 2007, the prosecuting authorities investigated the waste imported by S company. This investigation showed that imported coal ash includes 2.1 mg/kg of carcinogenic material, which is higher than that of designated waste. However, the prosecution did not have the authority under the existing regulations to impose any punishment on that company.

REFERENCES

Japanese

Kojima, Mitsukazu (2011), 'Economic development of developing countries and Basel convention', in A. Terazono (ed.), *The Export Situation and Proper Management Policy from the Perspective of Hazardous Waste Management, Fire Prevention, and Resource Recovery*, Ibaraki: National Institute for Environmental Studies.
Tsuruta, Jun and Aya Yoshida (2009), 'Review from the perspective of law', in A. Terazono (ed.), *The Export Situation and Proper Management Policy from the Perspective of Hazardous Waste Management, Fire Prevention, and Resource Recovery*, Ibaraki: National Institute for Environmental Studies.

Korean

Choi Byung-sung (2007), 'Hair test of residents nearby kiln', pp. 14–18.
Hangyere (2006), 'A large hole in the import and export management of hazardous wastes', Nov. 28.
Kim Hong-gyun (2007), *Environmental Law-case Study*, Seoul: Hongmun Publishing Company.
Kim Yong-won (2008), 'A study on the effectiveness of export-import control of wastes in Korea', *Journal of Korea Research Society for Customs*, **9** (3).
Korea Chemical Research and Development Institute (KCRDI) (2007), 'Analysis material on hazardous dust on Young-il village'.
Lee Seung-hi, Min Dal-gi and Sin Sung-gyung (2008), 'A study on relationship between designated waste and transboundary waste', *Journal of Korea Society of Waste Management*, **25** (4), 295–301.
Min Dal-gi (2009), *The Permission Criteria of Imports Waste Considering Nature and Status*, Inchon: Gachon University.
MOE (Ministry of Environment) (2004), *Commentary on Act of Waste Management*.
MOE (2008a), Environmental white paper.

MOE (2008b), *Guideline on Import and Export Declaration System.*
MOE (2008c), Submitted material to the Environment and Labor Committee for national inspection.
MOE (2009), *On Waste Export and Import Declaration System.*
MOE (2010), *A Guide on Discrimination of Transboundary Waste.*
MOE (2011), *Current Situation of Waste Import/export of Korea.*
The Board of Audit and Inspection of Korea (BAI) (2009), 'The actual state on cement hazard and utilization of waste in kilns'.

6. Challenges in the waste management system in the era of globalization: the case of the Philippines

Vella Atienza

INTRODUCTION

Like other developing countries, waste management has become a major problem in the Philippines. This situation has become more complex as it not only concerns the environmental component but it also includes economic, political and social considerations. In addition, with the era of globalization and modernization the composition of waste has become more diverse and its volume has been rapidly increasing in the past decades. Given the complexity of waste composition and the limited capability of each country to manage its own waste, the trading of waste and products from waste have become more internationalized. Although this international waste trading has brought some positive benefits especially in terms of economic aspect, it has also caused some environmental and health problems due to improper handling of waste and the lack, or the weak implementation, of policies to properly control the movement of waste and products from waste among countries. Thus, it is important to study the current policies of each country and how these policies can be improved to address these concerns on waste and how collaboration among countries and regions can be strengthened.

Given this growing problem of waste, many governments have included waste management programs in their priority agenda. In this chapter, the discussion focuses on the Philippines and its policies and program on managing waste, particularly the country's most recent national policy on waste, the Ecological Solid Waste Management Act of 2000, also known as the Philippines Republic Act 9003 (RA 9003). Specifically, this chapter emphasizes the need to include or strengthen the international aspect of managing waste to these national policies, especially as the trading of waste has been affected by both domestic and international factors and as the market for waste has also become more internationalized. The first section presents the salient features of the RA 9003 and the mandates to promote proper waste management and recycling. Other related policies on regulating imports of waste will also be discussed and the need for inclusion of the international aspect of waste trading for a sustainable waste recycling industry. The second section describes the significance of the proper collection system to facilitate recovery of waste and its relationship to the international trading of waste materials and products made from waste. The third section presents the flow of recyclables from the waste generators to waste consolidators, processors, manufacturers and exporters to illustrate why there is a need to recognize the significance of including the international aspect of waste trading. To give some examples of the possible forums in which the Philippines can take opportunities to discuss economic integration in the aspect of international waste trading, the fourth section discusses the Philippines-Japan Economic Partnership Agreement (PJEPA) and other free trade agreements (FTAs). Some issues and concerns on these bilateral and multilateral relations will also be presented. The last section is the summary and conclusion, emphasizing the need to include the international aspect of waste trading to strengthen the country's waste recycling industries.

6.1   THE ECOLOGICAL SOLID WASTE MANAGEMENT ACT AND OTHER WASTE RELATED POLICIES

The increasing population has contributed to the growing problem of waste in the Philippines. The country's population continues to increase at an accelerating rate and it is the "sixteenth most populous, out of more than 190 countries" (Magalona and Malayang 2001: 65). The Philippines' population exhibited a huge increase from 27 million in the 1960s to 88.57 million in 2007 (Espaldon and Baltazar 2004; NSO 2011). The annual population growth rate was 2.04 percent for the period 2000–07 (NSO 2011). Aside from the increasing population, rapid urbanization also contributes to the country's problem of waste. The unit generation rate of solid waste in the Philippines ranges between 0.30 and 0.70kg per capita per day for rural and urban communities respectively. As the population growth continues and given the stage in the socio-economic development in the country, it is estimated that the waste generation will also increase rapidly within the next few years (NSWMC 2005). Just as in other developing countries, the limited resources and inefficient waste collection system as well as political, economic and social considerations aggravate the problems of waste. The prevalence of open dumpsites in the Philippines has also caused not only environmental and health problems but also the very lives of those living near the area. To cite an example, the Payatas open dumpsite tragedy in July 2000 killed about 200 people, when the heavy rains caused a landslide of the former disposal site.

6.1.1 The Implementation of the Philippines Republic Act 9003, also known as the Ecological Solid Waste Management Act (ESWMA) of 2000

In response to the critical condition of the solid waste management problem and the threat it poses to the environment and human health if it remains unsolved, the Philippine government enacted the Republic Act 9003 on January 26, 2001. Unlike previous environmental policies that used a piecemeal approach, the Ecological Solid Waste Management Act of 2000 takes a holistic approach to the problems of solid waste management. It declares the intention of the state to adopt a systematic, comprehensive and ecological solid waste management program that will ensure the protection of public health and environment (Republic of the Philippines, RA 9003, 2001, Article 1, Section 2).

The National Solid Waste Management Commission (NSWMC) was created under the Office of the President, primarily to prescribe policies to attain the objectives of the Act and to oversee the overall implementation of the solid waste management programs. The NSWMC is chaired by the Secretary of DENR with members from 14 government sectors and three members from the private sector. The private sector includes representatives from NGOs, the recycling industry, and from the manufacturing and packaging industries (Republic of the Philippines, RA 9003).

The salient features of RA 9003 include the following: the primary role of the local government units (LGUs) in the implementation (Section 10); the mandatory closure of all dumpsites (Section 37) and the construction of sanitary landfill (SLF) as a final disposal site for residual wastes, in accordance with the criteria provided by the Act (Sections 40, 41, and 42); the participation of all sectors (Section 5q) such as the individual citizens, the recycling industry, manufacturing or packaging industries in the implementation of the Act (Sections 11, 12, 52); and the mandated 25 percent waste diversion of their generated waste within five years after the implementation of the Act through the promotion of composting, re-use and recycling activities. It further states that the reduction should be increased every three years (Section 20).

In addition, the National Ecology Center has to be established under the Commission to provide consulting, information, training and networking services for the proper implementation of the Act (Section 7) including the creation of the national recycling network (Section 7c). To promote recycling, the Act also mandates the establishment of material recovery facility (MRF) in every barangay[1] or cluster of barangays (Section 32); inventory of existing markets for recyclable materials (Section 26); an eco-labeling program (Section 27); and the development of a recycling market (Section 31).

6.1.2   Other Related Policies to Regulate Import of Waste

Other waste related policies which regulate the importation of waste include the following: the Philippines Republic Act (RA) 6969; the RA 4653; and the Department of Environment and Natural Resources (DENR) Administrative Order (DAO) 94-28. Of these, RA 6969, or the Toxic Substances and Hazardous and Nuclear Waste Control Act of 1990 "regulates the importation, manufacture, processing, handling, storage, transportation, sale, distribution, use and disposal of chemical substances and mixtures in the Philippines including the entry, even in transit, as well as the storage and disposal of hazardous and nuclear waste into the country for whatever purpose" (Republic of the Philippines, RA 6969, 1990). RA 4653 is an "Act to safeguard the health of the people and main-tain the dignity of the nation by declaring it a national policy to prohibit the commercial importation of textile articles commonly known as used clothing and rags" (Republic of the Philippines, RA 4653, 1966). DAO 94-28, or the interim guidelines on the importation of recyclable materials containing hazardous wastes, cites that it allows the importation of the following recyclable materials: scrap metals (lead acid batteries and metal bearing sludge), solid plastic materials, electronic assemblies and scrap (DENR, 1994). However, all importation must follow the requirements and procedures cited in the Basel Convention such as the notification and consent between parties, and the stipulation that the wastes to be imported must have a definite receiving facility with the essential environmental permits and clearance.[2]

6.1.3   The Philippines' National Policies on Waste and the Concept of International Waste Trading

From the above mentioned national regulations on waste, it can be noticed that the aspect of the international trading (import and export) of waste and products from waste seems lacking. RA 9003 cited how waste can be properly collected, recovered and processed/treated, and residuals disposed of. Also, it mentioned activities to promote recycling and develop a market for recycled goods, but it did not clearly specify if the markets also include the global market and how the international competitiveness of the local industries can be strengthened. Except for the brief mention about the fiscal and non-fiscal incentives for importing vehicles, equipment and other materials to be used for collecting and recycling waste (Article 45), it seems that the aspect of international trading to strengthen the recycling industry is not clearly emphasized in the Act. On the other hand, RA 6969, RA 4653, and DA 94-28 only cited some regulations on regulating import of toxic and hazardous waste and the list of permitted and not permitted waste and recyclable materials in the country. But these laws failed to recognize how to facilitate import and export of other types of waste and recyclables, especially the non-hazardous materials, which have already been occurring in the country for the past decades.

The experience of the Kilus Foundation Multi-purpose Cooperative in Barangay Ugong, Pasig City in the Philippines is one of the good examples on how the availability of a market for recycled products, in this case made from doy-pack[3] bags, has helped to sustain recycling activities. Through their participation in international trade shows, they are able to establish a relationship with foreign markets, and they have been exporting their products to about 17 countries including England, the US, Japan and Australia, among others. This activity has been providing a livelihood for more than 200 families (Personal Interview with Gina Santos, August 26, 2009 as cited in Atienza, 2010).

Using the experience of the selected case studies,[4] the next two sections discuss the significance of including the international aspect of managing waste to the national policy especially as the recycling market for waste has also become more internationalized.

6.2   IMPORTANCE OF A PROPER WASTE COLLECTION SYSTEM TO FACILITATE RECOVERY OF WASTE AND ITS RELATIONSHIP TO INTERNATIONAL WASTE TRADING

6.2.1   Promotion of Waste Segregation at Source and Creation of Material Recovery Facilities

To promote the 3R (reduce, reuse, recycle), RA 9003 also mandates the segregation of solid waste at source (Section 21) and the creation of material recovery facilities (MRFs) in every barangay or cluster of barangays (Section 32). The barangay is responsible for the collection of the

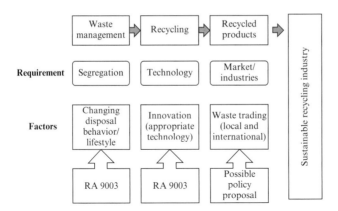

Source: Created by the author.

Figure 6.1   Factors needed for a sustainable recycling industry

segregated biodegradable and recyclable wastes, while the city or munici-
pality is responsible for the collection of non-recyclable and special wastes
(Section 10).

Research shows that a large percentage of waste generated in the
Philippines is made up of organics and recyclable waste. For example, in
Metro Manila the greater percentage of waste generated is composed of
biodegradable (about 45 percent) and recyclable waste (about 45 percent).
In addition, about 74.14 percent of this waste comes from households and
16.9 percent comes from commercial establishments (MMDA, 2007). This
means that if only households and commercial establishments could apply
segregation at source and practice recycling, it would be a great help not
only to extend the lifespan of disposal sites but would also save a huge
amount of money in the hauling services of waste and also give a potential
source of income for the communities and individuals depending on waste
for a livelihood. In addition, it would also provide environmental benefits
such as clean surroundings and at the same time it would reduce further
exploitation of natural resources by utilizing recyclables instead of using
virgin materials in industries.

Figure 6.1 shows that a proper waste management through segregation
is necessary to promote recycling; and recycling can be possible by the use
of appropriate or possible technology. But to sustain the behavior of prac-
ticing waste segregation and recycling, there should also be benefit from
this activity and this can be possible by transforming the recyclables into
products. Without the recycling industry that can translate these recycling
activities (or the recycled products) into a resource (or economic benefits),

the implementation of proper waste management may eventually die. Thus, there is a need to create markets for recycled products to assure the sustainability of the recycling industry. Most industries require a stable supply of waste materials for a continuous operation, and while some countries do not have facilities to process waste, other countries on the other hand have the facilities but lack the supply of waste materials. Thus there is a need to strengthen the trading of waste materials at both the local and international level. As the market for recyclables and the sources of waste materials for industries have become more internationalized, there is a need to strengthen international collaboration and partnerships among countries and regions.

Also, the proper regulation and control of the flow of waste materials must be strictly enforced to avoid the problems that may arise in this international waste trading activity. In addition, the government should also provide policy support to promote the recycled products and to encourage businesses to invest in this kind of operation by providing benefits such as tax reduction, etc.

6.2.2   The Three Case Studies: Quezon City, San Carlos City and San Fernando City

Using the three case studies, this section illustrates the importance of a proper waste collection system to facilitate effective recovery and recycling of waste and its relationship to international waste trading. Various stakeholders in the recycling and trading of waste will also be identified, as will the effect on their waste related activities of outside or international factors, such as the economic recession. The three study sites selected for the study are Quezon City (QC), in Metro Manila (or the National Capital Region), San Carlos City (SCC) in Negros Occidental, and San Fernando City (SFC) in La Union (Figure 6.2). Quezon City has the biggest population and the highest number of barangays. San Carlos City is second in land area but has the least waste generation per capita. San Fernando City is smaller in land area but more urbanized. It hosts major businesses, financial institutions, regional offices, universities and colleges in Region 1.

All of the three sites have been recognized for their "best practice" in solid waste management. These LGUs have a variety of formal and informal waste collection and recovery systems. The sites also have distinct characteristics that could exemplify what happens in different sorts of cities: those far from the big recyclers and traders such as San Carlos City; highly urbanized cities such as Quezon City, which has easy access to waste traders and recyclers; and urbanizing cities such as San Fernando

Source: Atienza et al. (2012).

Figure 6.2   The Philippines: location of the three study sites

City, which has "medium" access in terms of geographic distance to consolidators and traders. The three sites also have an existing sanitary landfill and the presence of informal waste sectors inside and outside the disposal facility.

6.2.3   Waste Management Collection and Recovery System

The collection system for SCC is managed by the LGUs; SFC has city-managed and cluster and barangay-managed systems; and QC is managed by private contractors with some barangay-managed systems like the Barangay Payatas, Barangay Holy Spirit and Barangay Bagumbuhay. SFC and QC provide subsidies to the barangays with their own collection system.

All three cities have closed their dumpsites and are presently using

Table 6.1   Recyclables percentage recovery in three SLFs

Items	San Carlos City	San Fernando City	Quezon City
1. Paper	15	6	3
2. Plastics	62	12	10
3. Scrap metal/tin cans	23	6	11
4. Bottles/cullet		74	76
5. Assorted materials		2	

Source: Atienza et al. (2012).

sanitary landfills as their disposal facilities. Both SCC and SFC landfills are managed by the city while QC SLF is privately owned and managed. These three cities have MRFs in their SLFs and all have also applied waste diversion/recovery activities such as composting activities and recovery processes for recyclables. MRFs are either barangay-managed, or privately managed such as by schools, institutions, subdivisions and commercial establishments, garbage haulers, and accredited junkshops. Other forms of recovery activities include waste markets and recycling collection events in collaboration by LGUs, private sectors (mall operators, accredited junkshops and recyclers), and NGOs. The common types of recyclables recovered from the three SLFs are composed of paper, plastics, scrap metal/tin cans, bottles/cullet and other assorted materials. Table 6.1 shows the percent recovery in the three SLFs.

It is also observed that there are waste materials that are not collected or are collected but just stored because of their toxic and hazardous components, lack of market, lesser value and bulkiness, etc. Some of the problematic wastes or those that are difficult to recover include plastic bags, Styrofoam, assorted plastic containers (especially bulky and hard), and busted lamps, among others. Given this condition, it can be seen that there is a need to fill up the "gap" in the current collection and recovery system as there are various wastes that are not collected or are difficult to recover. Thus, there is a need to secure the market for these waste materials in order to increase their value, which in turn will increase collection and recovery. For special waste, there is a need to find a destination (market) which has the capacity and facility to properly treat these materials and enhance recovery of valuable components.

In the next section, the flow of recyclables and how the waste trading activities have been affected by both domestic and international factors will be discussed.

6.3   FLOW OF WASTE RECYCLABLES AND OTHER FACTORS AFFECTING WASTE TRADING

6.3.1   Flow of Recyclables: Domestic and International

To illustrate further why there is a need to recognize the significance of including the international aspect of recycling and waste trading, this section shows the flow of recyclables and other factors affecting the recycling industry from both domestic and international aspects, based on the information gathered from the three case studies and interviews with consolidators, processors, and related waste industry associations. Figure 6.3 shows the flow of recyclables from the waste generators to waste consolidators, recyclers, processors, manufacturers and exporters. This figure shows the important role of junkshops as "middleman/intermediary" between waste generators and processors, etc., and their relationship with both local sources of waste and the international market for waste materials.

Table 6.2 shows the destinations of waste recyclables based on the

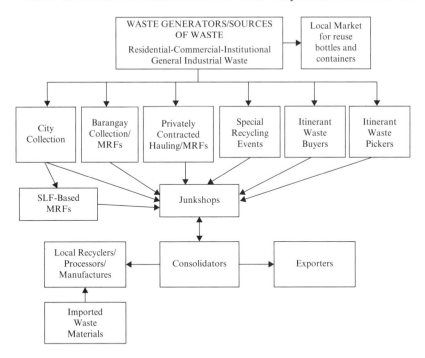

Source: Atienza et al. (2012).

Figure 6.3   Flow of waste recyclables

Table 6.2   Destinations of waste recyclables in the three study sites

Items	San Carlos City	San Fernando City	Quezon City
Paper	• Picked by CARPEL-QC – China	• Local consolidators: • TIPCO, Pampanga • Mangaldan and Dagupan Pangasinan	• CARPEL – for export to China • ILS Junkshop in Barangay Payatas
Plastics	• Bacolod City • Traders in Mandaue, Cebu for export	• SFC • Bauang, La Union • Mangaldan, Pangasian • Carmen, Rosales, Pangasinan • Valenzuela City	• Valenzuela City • Caloocan City
Bottles	• Negros Occidental – Tanduay Distillery • Cebu – San Miguel Bottling Company • For reuse in local markets	• SFC • La Union: • Sto Niño JS • Bauang, La Union • Sta Barbara, Pangasinan, La Tondeña Distillery	• San Miguel Bottling Company
Assorted metals	• Bacolod City • Cebu traders for export • Metro Manila traders for export or local smelters • Iligan City processor	• Poro Point, La Union for export by South Korean and Chinese consolidators • San Juan, LU • Dagupan City • Sison, Pangasinan	• Payatas consolidators/traders for local processors and export

Source: Atienza et al. (2012).

Table 6.3   Top ten destinations of the Philippines' export of waste plastics (HS 3915) in 2010

Country	Ton	%
1 China	89,677	68.4
2 Taiwan	10,507	8.0
3 Vietnam	8,873	6.8
4 Hong Kong	7,217	5.5
5 Malaysia	6,584	5.0
6 South Korea	2,571	2.0
7 Israel	1,596	1.2
8 Thailand	1,534	1.2
9 Tanzania	1,045	0.8
10 Japan	539	0.4
World	131,109	

Source: Extracted from World Trade Atlas data.

experience of the three case studies. Consolidators are either big junkshops, which act as consolidator for the small junkshops within the same barangay and/or neighboring communities; or consolidators/processors who purchase recyclables and engage in recycling activities; or consolidators who are primarily engaged in trading domestically or internationally. As shown in the table, most of the final consolidators and processors are located in Metro Cebu and Metro Manila.

Based on the interviews with the consolidators, most of the recyclables are exported to China. As shown in Tables 6.3–6.7, it is also observed and or confirmed that the destinations of common recyclables from the Philippines such as waste plastics, waste paper, iron scrap, copper scrap and aluminum scrap are China and other Asian countries.

Also, as shown in Figures 6.4–6.8, the Philippines is basically a net exporter of waste materials except for waste paper. But it can also be noticed that the gap between import and export of waste paper is becoming smaller.

In the case of second-hand goods like the cathode ray tube (CRT) TV, the Philippines is the highest importer of CRT TV from Japan (2008–11) among other Asian countries (Figure 6.9). Two reasons for this trend are the higher number of electronic repair shops in the Philippines to fix CRT TVs from Japan, and the fact that many individuals and families cannot afford to buy the more expensive type of TVs such as flat screen TV. Also, unlike other countries such as Indonesia, China, Malaysia and Thailand, the Philippines has had no regulation or prohibition on the import and export of second-hand TVs.

Table 6.4   Top ten destinations of the Philippines' export of copper scrap (HS 7404) in 2010

Country	Ton	%
1 China	9,411	44.5
2 Japan	4,294	20.3
3 Vietnam	2,031	9.6
4 South Korea	1,375	6.5
5 Indonesia	1,210	5.7
6 Taiwan	1,062	5.0
7 Singapore	601	2.8
8 Hong Kong	473	2.2
9 India	346	1.6
10 Italy	100	0.5
World	21,142	

Source: Extracted from World Trade Atlas data.

Table 6.5   Top ten destinations of the Philippines' export of waste paper (Hs 4707) in 2010

Country	Ton	%
1 Thailand	34,807	44.3
2 China	32,854	41.9
3 South Korea	5,691	7.3
4 Vietnam	4,379	5.6
5 Taiwan	355	0.5
6 Hong Kong	262	0.3
7 Indonesia	56	0.1
8 Canada	47	0.1
9 United States	23	0.0
10 Pacific Is. (Trust Terr.)	10	0.0
World	78,490	

Source: Extracted from World Trade Atlas data.

6.3.2   Domestic and International Factors Affecting the Trading of Waste

There are several factors affecting the trading of waste and scrap. At the national level, the government regulation on the promotion of waste segregation at source and the creation of material recovery facilities as mandated in RA 9003, and recent recovery activities such as recycling

Table 6.6   Top ten destinations of the Philippines' export of iron scrap (HS 7204) in 2010

Country	Ton	%
1 Taiwan	113,474	22.1
2 Indonesia	76,236	14.9
3 Thailand	62,864	12.3
4 Malaysia	47,222	9.2
5 South Korea	44,935	8.8
6 Singapore	43,873	8.6
7 China	42,042	8.2
8 Bangladesh	36,626	7.1
9 Japan	26,249	5.1
10 India	7,685	1.5
World	512,925	

Source: Extracted from World Trade Atlas data.

Table 6.7   Top ten destinations of the Philippines' export of aluminum scrap (HS 7602) in 2010

Country	Ton	%
1 Malaysia	8,580	41.4
2 South Korea	7,614	36.7
3 Japan	1,536	7.4
4 China	1,115	5.4
5 Singapore	468	2.3
6 Hong Kong	413	2.0
7 Thailand	387	1.9
8 Taiwan	234	1.1
9 Indonesia	199	1.0
10 United Kingdom	106	0.5
World	20,737	

Source: Extracted from World Trade Atlas data.

collection events and waste markets have facilitated the effective waste collection and recovery of recyclables. On the other hand, the recent move to ban the use of plastics in the country has had a negative impact on the plastic industry. As of January 2012, seven cities in Metro Manila banned the use of plastic bags for certain types of goods and Styrofoam food packaging while the rest of Metro Manila cities have similar pending

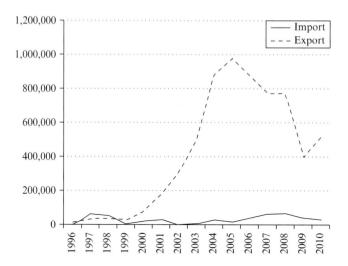

Source: Extracted from UN Comtrade data and World Trade Atlas data.

Figure 6.4   *Trends of import and export of iron scrap (HS 7204) in the Philippines (tons)*

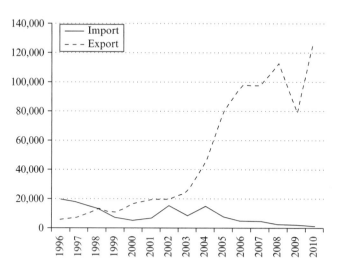

Source: Extracted from UN Comtrade data.

Figure 6.5   *Trends of import and export of waste plastic (HS 3915) in the Philippines (tons)*

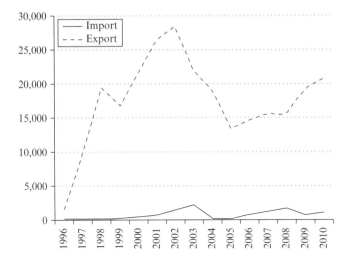

Note: Import data in 1997 has an irregularity, which is a sudden one time increase from United States. It is also not matched with US export data. The import data for 1997 was estimated, using US export data to the Philippines, instead of Philippines import data from the US.

Sources: Extracted from UN Comtrade data and World Trade Atlas data.

Figure 6.6   *Trends of import and export of aluminum scrap (HS 7602) in the Philippines (tons)*

ordinances. The Ecowaste Coalition listed more than 20 LGUs with plastic bans including those outside Metro Manila (Atienza et al. 2012). This has resulted in a shift to using paper bags and other reusable bags. It was reported there was about a 20 percent to 40 percent drop in the plastic bag manufacturing sector among the 300 manufactures (Lao 2011, as cited in Atienza et al. 2012).

Another factor that affects the recycling and trading activities is the archipelagic nature of the Philippines and the lack of infrastructure for efficient transportation. The country is an archipelago of 7,100 islands with a land area of 300,000 square kilometers. It is divided into three major geographical regions, Luzon, the Visayas, and Mindanao. It is composed of 17 regions, 81 provinces, 118 cities, 1,510 municipalities, and 41,995 barangays (Republic of the Philippines 2007). As of January 2011, out of 128 registered Treatment, Storage, Disposal (TSDs) for hazardous waste in the country, 95 are located in the National Capital Region (NCR) and other Luzon areas, and only 20 in Visayas and 13 in Mindanao (DENR-EMB, 2011). Since most of the big recycling industries

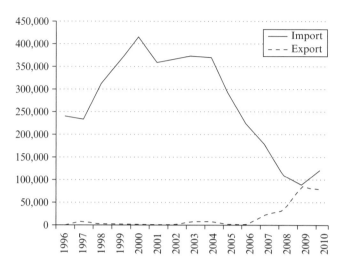

Source: Extracted from UN Comtrade data.

Figure 6.7   *Trends of import and export of waste paper (HS 4707) in the Philippines (tons)*

are located in Metro Manila and nearby provinces, the lack of recycling facilities in other regions restricts their recycling activities. The experience of barangay San Manuel in Puerto Princesa, Palawan (located at Region IV-B, about 567 km away from Manila) exemplifies this. Previously, the barangay had a Memorandum of Agreement (MOA) with the three junk-shops who collected its recyclables. In the year 2007, the barangay earned P34,000.00 for selling recyclables. But in the following year, the operation of junkshops stopped due to unavailability or a lesser number of trips for vessels/ships to transport waste from Palawan to Metro Manila. The sudden decrease in the cost of air transportation from Palawan to Manila and vice versa reduced the demand for sea transportation in the province, and this has affected the operation of waste trading in the area.

With regards to the buying and selling prices of recyclables, it is also observed that from the cases studied, MRFs in Quezon City's SLF (located at the NCR) have the highest while San Carlos City (located in the Visayas Region) has the lowest, primarily because of its inaccessibility to the big consolidators and processors and the higher cost of transportation, which affects the pricing of recyclables. For the sellers of recyclables, buying prices and accessibility dictate their decisions about where to sell their waste materials. It is also observed that recyclables with higher economic value are properly collected but those with lower or no value

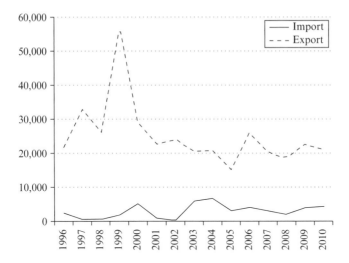

Note: Import data from 2003 to 2005 have an irregularity, which is a sudden one time increase from several countries. They are also not matched with exporting country data. The import data for 1997 was estimated, taking into account exporting country data and trend of import from the exporting countries.

Source: Extracted from UN Comtrade data and World Trade Atlas data.

Figure 6.8   *Trends of import and export of copper scrap (HS 7404) in the Philippines (tons)*

usually end up as residuals such as plastic waste (shopping bags, sachets, etc.) and Styrofoam as discussed earlier.

From the side of the formal industries, aside from the tedious requirement to secure licenses or permits, the high cost of investment and operation makes it difficult for them to compete with informal or non-registered recyclers/processors, who have strong purchasing power because they usually have much lower operating costs. Based on interviews with some registered TSDs, it seems that the government is very strict in implementing regulations with formal industries but that there is weak enforcement of the law for the non-registered or informal recyclers.

Aside from the domestic factors mentioned above, there are also international factors affecting the trading of waste materials in the Philippines, such as waste related international regulations and trends in the global market, among others. The demand and supply for recyclables and second-hand goods in the global market affect the local waste trading activities of the different stakeholders such as the informal waste sector, industries, processors, and consolidators. The stability of supply from the

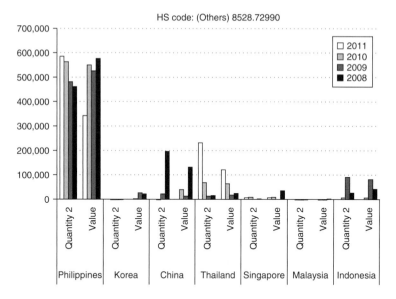

Source: Extracted from Trade Statistics of Japan data.

Figure 6.9   Japan's export of cathode ray tubes (CRT) TVs, 2008–11

local market dictates the local industry's decision on whether to import waste materials or not. For example, in the case of the plastic industry, the volume of materials must be assured or stable so that the operation of the facility is continuous. Thus, if they do not have enough materials from the local market, the industries tend to import waste resources.

For PET plastics, the high price offered by the foreign market encourages local consolidators to export plastic waste materials. As shown in Figure 6.5, there has been an increasing trend in the export of waste plastic in the Philippines except in 2008 during the economic recession. The high cost of electricity in the country (the highest in Asia) also affects the operation of recycling industries, especially electricity intensive industries such as plastic and steel; thus they have less competitive edge in the global market.

The economic recession in 2008 had a significant impact on the recycling and waste trading activities in the country, with a drastic fall in the prices of recyclables. As mentioned in Section 6.3.1, the Philippines is basically a net exporter of waste and scrap materials. Based on the interviews, many small recyclers, especially small junkshops, closed down or did not survive during the economic crisis. Bigger junkshops or consolidators were able to survive because they stockpiled their waste materials during the crisis and waited for the prices to increase before selling. However, informal waste pickers who depend primarily on waste picking for daily survival cannot afford to wait. Thus, they have no choice but to sell their recyclables even at lower prices.

The Philippines is also a signatory to international agreements including the Basel Convention on Control of Transboundary Movements of Hazardous Waste and their Disposal. However, it has not yet signed to ratify the Basel Ban Amendment. The details of the Basel Convention are discussed in Chapter 1. Other multilateral and bilateral agreements by the Philippines including the most recent, the Philippines–Japan Economic Partnership Agreement (PJEPA), and the next section discusses how this affects the trading of waste materials in the country.

6.4   THE PJEPA AND OTHER PHILIPPINES FTAS

To give examples of the possible forums in which the Philippines can take opportunities to discuss economic integration in international waste trading, this section discusses the Philippines–Japan Economic Partnership Agreement (PJEPA) and other free trade agreements (FTAs). Some issues and concerns arising from these bilateral and multilateral relations will also be presented.

6.4.1   The Philippines–Japan Economic Partnership Agreement

The Philippines–Japan Economic Partnership Agreement (PJEPA) is the first bilateral agreement by the Philippines and is also a controversial one, particularly because of the inclusion of tariff elimination for wastes, including hazardous wastes, by both Japan and the Philippines. Since both countries are signatory to the Basel Convention, they state that they will respect the regulations set by the convention particularly on the movement of hazardous waste. The two countries also claim that this agreement will be beneficial to both countries as it will facilitate trade and investment; however, many members of environmental non-government organizations (NGOs), not only in the Philippines but also international NGOs such as Greenpeace International, Basel Action Network (BAN) and Global Alliance Against Incinerators (GAIA), among others, strongly oppose such agreement and believe that it will only encourage or legalize Japan's use of the Philippines as a dumping site for its waste. The various environmental and health hazards if these wastes are not properly handled affect the position of the NGOs towards PJEPA, considering the various incidents of illegally transported hazardous waste in the guise of recyclables and second-hand goods in the past (as discussed in Chapter 9).

Based on interview, from the government side it is argued that the NGOs' claim (that the Philippines could become a dumping site for Japan's waste following tariff elimination) is not true. Just as for other goods, there is a control mechanism for the import of waste, including hazardous waste, which is the issuance of the import clearance certificate (ICC) as also stated by the Basel Convention. However, the question here is, does the Philippine government have enough capacity and facility to strictly monitor the entry of waste into the country and what is the assurance that this control mechanism could really be effective? It seems that the main issue here is the enforcement problem. Thus, even if the tariff elimination for waste was excluded from the Agreement, how could the public be assured that there would be no illegal entry of hazardous waste to the country if the enforcement itself is in question? Therefore, the element of trust between the government and the public, and how this trust can be established, also plays a role in this argument.

The removal of tariff is just one aspect of the PJEPA agreement. With regards to investment, the government also claims that PJEPA will facilitate more investment in the country and this would create more jobs. However, the increase of foreign investors in the Philippines may also have a negative impact on small industries which do not have enough capital to compete with foreign industries. This situation may also have the same effect in the case of the waste recycling industries as there are many small and medium recyclers in the country. This poses a challenge to the government to provide technology support and training for these small industries to enhance their capacity to improve their operations and compete with the global market.

Another issue in connection with the argument of environmental NGOs about the impact of the import of hazardous waste to the Philippines, is that, on the other hand, the Philippines is also exporting hazardous waste. To cite an example, it was reported that in 2009 DENR issued import clearance for electronic waste and scrap from Hong Kong (350,000 pieces), South Korea (50,000 pieces and 2,456 boxes), Seoul, South Korea (10,000 sets), United States of America (USA) (9,000 pieces), Thailand (1,100 tons), and Japan (22 pieces); and also export clearance to Thailand (134,000 kilograms), Japan (1,000 metric tons and 2,000 tons); Belgium (10,500 kilograms), and Singapore (800 tons) (Sañez, 2012).

This chapter shows both sides of the argument between government and the NGOs. At the same time it shows that although there is a risk involved in the international trading of waste, nevertheless, based on the existing global economy, this activity or movement of waste is inevitable. Thus, the best thing to do is to strengthen the international collaboration to avoid or prevent adverse incidents like those that have happened in the past, and to strengthen partnerships among countries so that both importers and exporters can benefit from this trade.

6.4.2 Other Multilateral Agreements

The Philippine government has entered into various multilateral and bilateral free trade agreements (FTAs) in the past two decades such as the ASEAN FTA (AFTA), ASEAN China FTA (ACFTA), ASEAN Korea FTA (AKFTA), ASEAN–Japan Comprehensive Economic Partnership Agreement (AJCEPA), ASEAN–Australia–New Zealand FTA (AANZFTA), ASEAN–India FTA (AIFTA), and the PJEPA. Free trade causes the influx of cheap imported goods to the country, and this may kill local industries, especially small industries, due to stiff competition from foreign companies. On the other hand, free trade also provides opportunities to enhance the export market of recyclables from the Philippines to other countries, and the higher buying price from foreign countries could also benefit local waste consolidators/recyclers. However, this condition also creates a negative impact on other local industries because of decreasing supply of waste materials from the local sources, due to strong competition with foreign markets.

As shown in Figure 6.7, the Philippines is a net importer of waste paper; thus the increasing trend in the export of waste paper recently created a negative impact on the local paper manufacturers in the country. According to the Philippine Pulp and Paper Manufacturers Association (PULPAPEL), representing the paper industry, this trend threatened the security of raw material supply in the country. Thus, in 2008 they called on the government to ban the export of waste paper. Bataan 2020, one of the top paper manufacturers in the Philippines, stated that more than 1,000 tons of waste paper were regularly shipped out to China and Hong Kong, that this volume was increasing, and that there were also buying inquiries from Vietnam, Indonesia and Australia.[5]

6.5 SUMMARY AND CONCLUSION

This chapter reviews the current waste management system in the Philippines in its effort to address the growing problems of waste, and the related political, economic and other social issues. Based on the discussion, it shows that proper waste management through strict implementation of waste segregation at source promotes recycling and enhances the recovery of recyclables, which in turn promotes waste trading activities. It is also observed from the cases studied that geographical location affects the percentage recovery of recyclables. This is because accessibility to the big consolidators and processors and the cost of transportation affect the effectiveness of recycling and trading activities.

Based on the cases discussed, it is shown that recyclables with higher economic value are properly collected, such as paper, PET bottles, scrap metals, etc. However, there are also other waste materials which are considered problematic or which are more difficult to recover due to toxic and hazardous components, lack of market, lesser value and bulkiness, etc. Thus, there is a need to secure the market for these waste materials to increase their value, which in turn would increase the collection and recovery. For special waste, there is a need to find a destination (market) which has the capacity and facility to properly treat these materials and enhance recovery of valuable components. This market or destination can be either within the country or in other countries, but proper control should be strictly enforced to avoid a negative impact on the environment and human health.

As discussed in this chapter, one of the major constraints for waste trading activities in the far provinces is the cost of transportation due to geographical location. One recommendation is that the government considers designating some of the country's ports as "Recycle Ports," especially in remote areas which have difficulty transporting recyclables to Manila but may find a market in neighboring countries such as China. For example, in Japan the government, with the support of the Ministry of Land, Infrastructure and Transport's Ports and Harbours Bureau, has designated 18 ports as "Recycle Ports" (MILT, 2012).

Japan also supplies an example of how to deal with one of the waste identified as problematic in the study – Styrofoam – due to its lack of, or limited, market. The Tokyo Metropolitan Central Wholesale Market in Tsukiji generates more than 50,000 plastic foam or Styrofoam boxes every day for use in packing fish, and had difficulty in disposing of this material, just like the Philippines. But compactors are used at the Tsukiji market to squeeze the sheets to one-fifth of their original size, and they are then shipped overseas for recycling in China. The plastic foam is crushed into fragments and processed to make the black frames for video-cassette tapes, compact disk cushions, etc. These recycled products are then exported back to Japan and other countries (*Asahi Shimbun*, April 2, 2004). This could also be one of the possible solutions to address the problems of handling Styrofoam in the Philippines.

Results of the study also show that the recycling and trading activity in the country is affected not only by domestic factors such as national regulations, geographical location, and other internal issues as mentioned above, but also by international factors that create a positive or negative impact on waste recycling operations, such as demand and supply from the foreign market, the economic recession, and the Philippines' membership of various multilateral and bilateral agreements such as the PJEPA, among others.

As shown in Figure 6.3 (Flow of Recyclables), the movement of waste from one country to another is inevitable in this era of globalization, wherein the market for recyclables and the sources of waste materials for industries have become more internationalized. Thus, there is a need for the government to also include the international aspect of recycling and waste trading in its national policy to sustain the country's waste recycling industry. Also, there is a need to strengthen international collaboration and partnerships among countries and regions, especially with the neighboring Asian region, to increase global competitiveness.

However, as shown in this chapter, entering into bilateral agreements and other multilateral partnerships may also create some issues, as shown by the arguments between the government and the environmental NGOs over the ratification of the PJEPA. Thus, transparency and active participation among various stakeholders should be promoted to assure both parties that the welfare of both nations is protected and that this activity will be beneficial to both countries. Also, as international collaboration may increase the country's competitiveness in the global market, it may also create a negative impact on the local small and medium industries that do not have enough capital and skills to compete with foreign industries. Thus, the government should provide support for these small industries by enabling them to improve their operations and to compete with the global market.

ACKNOWLEDGMENTS

The author would like to thank the following for their significant contributions in the data gathering and for generously sharing their insights on waste management, particularly on recycling and waste trading: Mr Michikazu Kojima and Ms Etsuyo Michida (IDE-JETRO); Ms Lizette Cardenas, Ms Eileen Sison and Mr Victor Arvin Sison (SWAPP); and the key informants and respondents in this study from the government, private industries and associations, NGOs, and members of the informal waste sector in the Philippines.

NOTES

1. Barangay is a smallest political unit in the Philippines.
2. Sañez, Geri Geronimo. 2012. "E-waste issues in the Philipines", Paper presented during the 8th NIES Workshop on E-wastes, Quezon City, Philippines, 25 January 2012.
3. Doy-pack bags are made of triple laminated aluminum cellofoil (Kilus 2012).
4. The data and information for these case studies are primarily based on the results of the IDE-SWAPP (Solid Waste Management Association of the Philippines) joint research project ("sub-project") led by the author, as part of the "Economic Integration and Recycling in Asia" Research Project.
5. Bataan 2020, Inc. (2012), http://www.bataan2020.net/newsroom.htm.

REFERENCES

Asahi Shimbun (2004), 'What a waste: but China has a global edge', 2 April http://www.asahi.com/english/asianet/hatsu/eng_hatsu040402a.html (accessed 19 February 2012).

Atienza, Vella (2010), 'Benefits and strategies to improve the condition of the informal waste sector in the Philippines', United Nations Center for Regional Development (UNCRD)'s Regional Development Dialogue (RDD), **XXXI** (2) (Autumn), 62–82.

Atienza, V., E. Sison and L. Cardenas (2012), 'Review of the Philippines' waste segregation and collection system and the trading of recyclables', a joint research project by the Institute of Developing Economies – Japan External Trade Organization (IDE-JETRO) and the Solid Waste Management Association of the Philippines (SWAPP), IDE-JETRO Joint Research Program Series No. 158.

Bataan 2020, Inc. (2012), 'Ban on waste paper: threat to security of material supply in RP', http://www.bataan2020.net/newsroom.htm (accessed 1 February 2012) (original article taken from the *Philippine Daily Inquirer*, 15 April 2008).

DENR (Department of Environment and Natural Resources) (1994), Administrative Order (DAO) 94-28, *The Interim Guidelines on the Importation of Recyclable Materials Containing Hazardous Wastes.*

DENR-Environmental Management Bureau (DENR-EMB) (2011), official website of the DENR-EMB, http://www.emb.gov.ph (accessed 21 February 2011).

Espaldon, M.V.O. and M.R.M. Baltazar (eds) (2004), *Participatory Natural Resource Management for Sustainable Agriculture*, Quezon City, Philippines: Department of Geography, College of Social Sciences and Philosophy, University of the Philippines.

KILUS (2012), official website of the KILUS Foundation, http://www.kilus.org (accessed 19 February 2012).

Lao, Crispian (2011), 'Recycling and proper management of plastics', paper presented during the SWAPPCON 2011, Cebu City, Philippines, November.

Magalona, Merlin and Ben S. Malayang (2001), 'Environmental governance in the Philippines. Philippineas', Proceedings of the International Symposium on Environmental Governance in Asia, Tokyo, Japan, Sophia University, 9 March 2000.

Metro Manila Development Authority (MMDA) (2007), 'Door-to-door garbage collection in Metro Manila', Paper presented during the SWAPP Con 2007, Baguio City, 19–23 November.
Ministry of Land, Infrastructure, Transport and Tourism (MILT) (2012), official website of Japan's MILT, http://www.mlit.go.jp/en/index.html (accessed 19 February 2012).
National Solid Waste Management Commission (NSWMC) (2005), *Technical Guidelines on Solid Waste Disposal Design and Operation*, Quezon City, Philippines: NSWMC.
National Statistics Office (NSO) (2011), official website of the NSO, http://www.census.gov.ph (accessed 8 March 2011).
Republic of the Philippines, RA 4653 (1966), *Act to Safeguard the Health of the People and Maintain the Dignity of the Nation by Declaring it a National Policy to Prohibit the Commercial Importation of Textile Articles Commonly Known as Used Clothing and Rags.*
Republic of the Philippines, RA 6969 (1990), *Toxic Substances and Hazardous and Nuclear Waste Control Act of 1990*, http://www.gov.ph/1990/10/26/republic-act-no-6969/ (accessed 11 February 2012).
Republic of the Philippines, RA 9003 (2001), *Ecological Solid Waste Management Act of 2000*, http://eia.emb.gov.ph/nswmc (accessed 25 February 2010).
Republic of the Philippines. (2007), official website of the Republic of the Philippines, http://www.gov.ph/aboutphil/general.asp (accessed 24 May 2007).
Sañez, Geri Geronimo (2012), 'E-waste issues in the Philippines', Paper presented during the 8th NIES Workshop on E-wastes, Quezon City, Philippines, 25 January.
Santos, Gina (2009), 'Personal interview', Ugong, Pasig City, Philippines, 26 August.
Trade Statistics of Japan (2012), http://www.customs.go.jp/toukei/info/index_e.htm (accessed 10 February).
United Nations Commodity Trade Statistics Database (UN Comtrade) (2012), http://comtrade.un.org/db/ (accessed 6 January 2012).
World Trade Atlas Database (2012) (accessed 16 February 2012).

7. Japanese implementation of the Basel Convention and its problems

Jun Tsuruta

INTRODUCTION

The purpose of this chapter is to review the measures taken by the Government of Japan to implement the Basel Convention on the Control of Transboundary Movements of Hazardous Wastes and their Disposal, which was adopted in March 1989 for reducing and controlling transboundary movements of hazardous wastes in order to protect human health and the environment. Section 7.1 deals with the drafting history of the Basel Convention and its contents, in particular, its basic principles, its obligations, and its procedure of prior informed consent for importing/exporting hazardous wastes. Section 7.2 deals with the measures taken by the Government of Japan to implement the Basel Convention in Japan, especially establishing the Japanese domestic act for implementing the Basel Convention ("Japanese Basel Act") and regulating export of hazardous wastes from Japan to other States. Section 7.3 deals with the Japanese Basel Act's enforcement and its problems regarding export regulations on hazardous wastes. Finally, Section 7.4 deals with the measures taken by the Government of Japan to make export regulations on hazardous wastes more effective.

7.1 THE BASEL CONVENTION FOR ENSURING THE APPROPRIATE TRANSBOUNDARY MOVEMENTS OF HAZARDOUS WASTES

Transboundary movements of hazardous wastes have been a common occurrence since the 1970s, primarily between the United States and European countries. In the early 1980s, however, the United States and European countries began to export their hazardous wastes to developing countries in Africa and other regions where environmental regulations and their enforcement were less developed or lacking, in order to dispose of and recycle waste. This practice often resulted in pollution that affected human health and the environment in developing countries, as the final disposal and recycling of waste can be harmful. Further, in the absence of adequate safeguards, recycling and recovery operations especially have the possibility of causing greater health danger given the higher level of workers' exposure and handling.

During the 1980s there was increasing concern about industrialized countries exporting hazardous wastes to developing countries that lacked the legal, administrative and technological resources to manage the disposing or recycling of such wastes appropriately. Also, several prominent cases of illegal or mismanaged transboundary movements of such wastes occurred during this period. International legal regulation of transboundary movements of such wastes began with efforts in the Organisation for Economic Cooperation and Development (OECD) and the European Community to improve these problematic circumstances in the early 1980s.

The United Nations Environment Program (UNEP) also began serious consideration of the issues concerning generation and transportation of hazardous wastes in the early 1980s. In 1985, the non-binding Cairo Guidelines and Principles for the Environmentally Sound Management of Hazardous Wastes were published and then approved by UNEP in June 1987. In addition to the declared aim of ensuring human health and the environment against the adverse effects of hazardous wastes, the Cairo Guidelines also adopted the procedure of "prior informed consent" by all of the States concerned with the transboundary movements of such wastes. Also in June 1987, UNEP's Governing Council requested that the Executive Director prepare a global legal instrument to control the transboundary movements of wastes. The OECD had already begun working on a draft international agreement, and so the initial efforts of the UNEP group were based in large part on the OECD draft, which did not envision a complete ban on transboundary movements of such wastes but rather the establishment of appropriate transboundary movements.

The Basel Convention on the Control of Transboundary Movements of Hazardous Wastes and their Disposal ("Basel Convention") was adopted in March 1989, and came into force in May 1992. It came into effect in Japan in September 1993. As of April 2012, 179 States and the European Community became parties to the Convention. Now, the Basel Convention plays an important role in the prevention and elimination of the worst forms of "toxic dumping" by industrialized countries in developing countries in environmentally unsound manners, and its Secretariat provides advice and circulates the relevant information to parties to the Convention.

The objectives of the Basel Convention are to minimize the generation of "hazardous wastes" or "other wastes" (hereafter, collectively referred to simply as "hazardous wastes")[1] and to reduce and control transboundary movements of such wastes in order to protect human health and the environment. To achieve these objectives, the Convention stipulates several general legal obligations: that the export of such wastes are prohibited to Antarctica, to parties having banned the import of such wastes, and to States not party to the Basel Convention unless they conclude an agreement that is as stringent as the Convention, under Paragraphs 1, 5, and 6, Article 4 of the Convention. Parties may, however, enter into bilateral or multilateral agreements on managing hazardous wastes with other parties or with non-parties, provided that such agreements are "not less environmentally sound" than the Basel Convention, in accordance with Article 11 of the Convention.

Also, the Basel Convention does not, in principle, aim to ban transboundary movements of hazardous wastes, but rather to ensure that their movements do not endanger human health and the environment of the developing countries importing them.[2] In other words, their transboundary movements from the State of their generation to any other State may be permitted only under conditions that do not endanger human health and the environment, and only under conditions in conformity with the provisions of the Basel Convention.[3]

To realize this principle, the Basel Convention established the procedure of prior informed consent (PIC). This mechanism requires the prospective exporting States to submit a prior written notice detailing the export plan regarding hazardous wastes to the competent authorities in the prospective States of import and transit, and to obtain their written consent. The transboundary movement may proceed only when all of the States concerned have given their written consent. When written consents are granted by all the States concerned, the prospective exporting State gives permission to the exporters to commence the export procedures. On the other hand, if written consents are not granted by them, the exporting State may not give permission to the exporter, or is obligated to suspend such export.

A number of provisions of the Basel Convention deal with inappropriate cases where transboundary movements of hazardous wastes have been carried out illegally, or cannot be completed as foreseen. In such cases, the Convention attributes the responsibility under international law to the States concerned and imposes on them the obligation to ensure safe disposal of such wastes, for example, by re-import into the State of export by the exporter or the generator of such wastes or, if necessary, by the State of export itself. Transboundary movement of such wastes not in accordance with the PIC procedure is defined as "illegal traffic" by Paragraph 1, Article 9 of the Basel Convention,[4] and parties to the Convention are obliged to establish appropriate national domestic laws and regulations to prevent and punish such illegal traffic under Paragraph 5, Article 9 of the Convention.[5] More generally, the Basel Convention obligates parties to take measures such as establishing domestic laws and regulations under Paragraph 4, Article 4, stipulating that "Each Party shall take appropriate legal, administrative and other measures to implement and enforce the provisions of this Convention, including measures to prevent and punish conduct in contravention of the Convention." These obligations were stipulated under the Convention because each party's domestic laws and regulations were partly responsible for creating the problematic situation of transboundary movements of such wastes, and so, in order to improve such situation, it was necessary to harmonize them with international standards.

As described above, the purpose and content of the Basel Convention is not to prevent transboundary movements of hazardous wastes but to ensure their appropriate movements. However, even after the Basel Convention came into force, there has been no end to illegal transboundary traffic of such wastes. For example, there have been many cases in which such wastes declared by the exporter to be "recyclable resources" or "second-hand items" have not been exported in accordance with the PIC procedure, have been inspected and recognized not to be "recyclable resources" or "second-hand items" but "hazardous wastes" by border control authorities of the importing State, and have been shipped back to the exporting State.

7.2 THE JAPANESE BASEL ACT FOR IMPLEMENTING THE BASEL CONVENTION IN JAPAN

The prevention and punishment of "illegal traffic" of hazardous wastes, and the attribution of responsibility derived from such illegal traffic to the exporting State when illegal traffic occurs, was one of the most contentious issues during the negotiations of the Basel Convention in the late 1980s. This problem remains on the agenda as one of the most difficult tasks for implementing the Convention in practice. The Convention requires Parties to establish national legislation to prevent and punish illegal traffic. Of course, it is not easy at all to prevent, identify, detect and reduce illegal traffic, which is a phenomenon of considerable complexity and magnitude. Illegal traffic occurs in many forms, and elaborate schemes are devised to circumvent national domestic and international regulation. The most promising approach is probably to continue to strengthen national domestic regulation and its enforcement by border control authorities, such as customs, quarantine, and maritime police.

In December 1990, the Central Public Nuisance Council of the Government of Japan recommended that the Government of Japan ratify the Basel Convention as soon as possible and establish a legal system for regulating movements of hazardous wastes. In accordance with this recommendation, the Government of Japan, before ratifying the Convention, began studying measures to implement the Convention, with a goal of early accession to the Convention in the fall of 1991. The Government of Japan then submitted to the Diet the "Draft Act on the Control of Export, Import and Others of Specified Hazardous Wastes and Other Wastes" that would implement the Convention in Japan. In January 1992, the Industrial Structure Council of the Government of Japan also recommended a measure that was basically the same. The Draft Act was passed by the 125th extraordinary session of the Diet in December 1992, and the "Act on the Control of Export, Import and Others of Specified Hazardous Wastes and Other Wastes" (Japanese Basel Act) was enacted as Act No. 108 of 1992 in December 1992 and took effect in December 1993. In addition, the Government of Japan amended the "Act on Waste Management and Public Cleansing" (Act No. 137 of 1969, the Waste Management Act) in order to implement the Basel Convention in Japan. The import to and export from Japan of hazardous wastes are regulated by this Act and the Japanese Basel Act. The Basel Convention itself came into force in May 1992 and came into effect in Japan in September 1993, the 90th day after the date of deposit with the Secretary-General of the United Nations by the Government of Japan of its instrument of ratification in accordance with Paragraph 2, Article 25 of the Convention.

In general, concerning measures for implementing international treaties in Japan, the provision of Paragraph 2, Article 98 of the Constitution of Japan, which states that "treaties concluded by Japan and established laws of nations shall be faithfully observed," is interpreted as recognizing that treaties concluded by Japan have the highest nature in the Japanese legal system, and so that they are generally incorporated into the Japanese legal system and in themselves are legally effective in its system, even without establishing domestic laws and regulations corresponding to rights and obligations stipulated by treaties. Therefore, establishing laws and regulations for implementing treaties is only one of the means to ensure their implementation in Japan, for example, when it is not possible for administrative and judicial agencies to directly apply or enforce provisions of treaties.

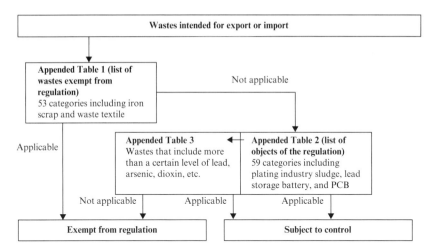

Figure 7.1   *Process for identification of the "specified hazardous wastes"*
              in the Notice to the Japanese Basel Act

The purpose of the Japanese Basel Act is to ensure accurate and smooth implementation of the Basel Convention by taking measures such as the regulation of import and export of "the specified hazardous wastes" ("materials contained in Annex I to the Convention that have any of the hazardous characteristics contained in Annex III to the Convention" in Paragraph 1(i)(a), Article 2 of the Japanese Basel Act) and "other wastes" (hereafter, collectively referred to simply as "specified hazardous wastes") subject to the regulation under the Convention, and thereby to contribute to preservation of human health and the environment (Article 1 of the Japanese Basel Act stipulates the purpose of the Act, which is provided in Appendix 7.1).

The definition of the specified hazardous wastes under the Japanese Basel Act is basically identical to the hazardous wastes under the Basel Convention, because the Japanese Basel Act defines the specified hazardous wastes directly by quoting Annex I, II, III, and IV to the Basel Convention. Annex I stipulates "categories of wastes to be controlled," Annex II does "categories of wastes requiring special consideration," Annex III contains a "list of hazardous characteristics," and Annex IV covers "disposal operations" (Article 2 of the Act stipulates the definition of the specified hazardous wastes as given in Appendix 7.2). Furthermore, the specified hazardous wastes under the Japanese Basel Act are clarified and contained in the Notification enacted in 1998 in accordance with Paragraph 1(1)(i), Article 2 of the Japanese Basel Act (Figure 7.1).

The specified hazardous wastes under the Japanese Basel Act overlap directly what the Government of Japan interpreted as the materials regulated by the Basel Convention. In this regard, the Government stresses "continuity" from the subjects regulated by the Basel Convention to those regulated by the Japanese Basel Act, by stating that "because the specified hazardous wastes as the materials regulated by the Basel Act are defined by quoting Annexes to the Basel Convention, if Annex I of the Convention . . . is amended, the content of the specified hazardous wastes are also automatically changed" (Ministry of Environment, the Government of Japan. Suishitsu-hozen-kyoku, Haikibutsu-mondai-kenkyukai (Water Quality Bureau, Hazardous Waste Study Group) (ed.) 1993, p. 126).

The Japanese Basel Act incorporates the procedures for the import and export regulations adopted by the Foreign Exchange and Foreign Trade Act (Act No. 288 of 1949) (hereafter, "Foreign Exchange Act"). Therefore, the relevant regulations and penalties stipulated in the Foreign Exchange Act shall apply *mutatis mutandis* to the import to and export from Japan of the specified hazardous wastes regulated by the Japanese Basel Act.

The procedure for exporting the specified hazardous wastes stipulated by Article 4 of the Japanese Basel Act is as follows: (1) a person intending to export the specified hazardous wastes applies for approval to the Minister of Economy, Trade and Industry in accordance with the procedure for the export regulations stipulated by Paragraph 3, Article 48 of the Foreign Exchange Act,[6] (2) the Minister of Economy, Trade and Industry forwards a copy of the written application submitted by the above-mentioned person to the Minister of the Environment, if the destination of export is a region designated as requiring special attention in order to prevent environmental pollution which the disposal of the prospective exported specified hazardous wastes have the possibility to cause, (3) the Minister of the Environment sends a copy of the written application to the prospective importing State to confirm that the State has taken appropriate measures to prevent the above-mentioned environmental pollution concerning the disposal of the prospective exported specified hazardous wastes, (4) the Minister of the Environment then informs the Minister of Economy, Trade and Industry of the result of the confirmation, (5) the consent of the prospective importing State is notified to the Minister of Economy, Trade and Industry through the Minister of the Environment, and (6) the Minister of Economy, Trade and Industry approves the application in accordance with the procedure for the export regulations stipulated by Paragraph 3, Article 48 of the Foreign Exchange Act (Article 3 of the Act stipulates the procedure for exporting the specified hazardous wastes as given in Appendix 7.3).

Concerning the relationship of the above mentioned procedures (4), (5), and (6), the consent of the prospective importing State is interpreted to be a "requirement" for the approval of the application of exporting by the Minister of Economy, Trade and Industry, and so the Minister shall not approve the application of exporting without the confirmation by the Minister of the Environment mentioned above in the procedure (4).

Paragraph 4, Article 4 of the Basel Convention obliged the Parties to "take appropriate legal, administrative and other measures to implement and enforce the provisions of this Convention, including measures to prevent and punish conduct in contravention of the Convention." So the Japanese Basel Act stipulates (a) the export regulation procedure of the specified hazardous wastes by incorporating the procedure stipulated by Paragraph 3, Article 48 of the Foreign Exchange Act, and (b) punishments for making "false statement" in the application for approval of export of such wastes, "false reporting," or "evasion of entrance and inspection," etc. Regarding violating regulations of the abovementioned procedure (a), any person who has exported goods without obtaining approval from the Minister of Economy, Trade and Industry shall be punished by "imprisonment with work for not more than 5 years or a fine of not more than 5,000,000 JPY," or both, in accordance with Paragraph 1, Article 69(7) of the Foreign Exchange Act. Any person who had obtained approval from the Minister of Economy, Trade and Industry by means of "false application" of exporting the specified hazardous wastes shall be punished by imprisonment with labor for not more than 3 years, a fine of not more than 1,000,000 JPY, or both, in accordance with Paragraph 1, Article 70 of the Foreign Exchange Act. And, regarding the above-mentioned (b), any person who has made "false statement" in the application for approval of export of such wastes, "false reporting" or "evasion of entrance and inspection," etc. shall be punished by "imprisonment with work for not more than 6 months or a fine of not more than 500,000 JPY, or both," in accordance with Article 22 of the Japanese Basel Act.

The Japanese Basel Act, as well as the Basel Convention, is based on the principle of domestic disposal of the specified hazardous wastes and thus the Act intends to restrict the transboundary movements of such wastes to a minimum. In addition, the Notification enacted by the Ministry of Economy, Trade and Industry (METI) and the MOE, etc. in 1993 in accordance with Article 3 of the Japanese Basel Act stipulates the basic principles on the specified hazardous wastes, such as "minimization of import and export of the specified hazardous wastes," "restrain of its generation," "promotion of minimizing of its hazardousness," "promotion of its domestic disposal," "promotion of technological developments," and "periodic study of the possibility of minimizing of its export and the possibility of minimizing of its hazardousness," etc. On the other hand, the Foreign Exchange Act is based on the market mechanism of foreign transactions but exercises "control and coordination" if excessive pursuit of profit by a private company threatens to give rise to obstacles to the normal development of foreign transactions. However, such "control and coordination" shall be within "the minimum necessary level."

Therefore, the Japanese Basel Act applies *mutatis mutandis* to the import and export procedures stipulated by the Foreign Exchange Act, but it needs to be noted that the Basel Convention and the Japanese Basel Act take a different approach to international transactions from that taken by the Foreign Exchange Act.

7.3   ENFORCEMENT AND ITS PROBLEMS REGARDING THE EXPORT REGULATIONS ON THE SPECIFIED HAZARDOUS WASTES UNDER THE JAPANESE BASEL ACT

Despite establishing the Japanese Basel Act to implement the Basel Convention in Japan, an incident regarding illegal traffic of hazardous wastes from Japan to the Republic of the Philippines occurred in December 1999 (the so-called "Nisso incident"), which developed into a diplomatic issue between Japan and the Republic of the Philippines. In this incident, Nisso Co. Ltd, a Japanese industrial waste disposal company, exported to the Philippines 2,160 tons of industrial wastes containing "medical wastes," one of the specified hazardous wastes under the Japanese Basel Act, by means of making "false statement" in the application for approval of export of the cargo that the content of the cargo was "valuable resources, waste paper (assorted paper), mixed material (plastic)." Therefore, the above mentioned procedures from (2) to (5) for exporting the specified hazardous wastes under the Japanese Basel Act, in which the MOE should also be involved under the Act, were not taken at all, but the METI approved Nisso's application for export of the cargo containing the specified hazardous wastes, not in accordance with the PIC procedure provided by the Basel Convention, and did so without the prior consent of the Government of the Republic of the Philippines.

In this incident, the Government of Japan recognized that this export was illegal traffic of hazardous wastes under the Basel Convention, and issued the administrative orders that the exporter Nisso should take the measures to re-import the cargo in question to Japan and to dispose of it appropriately in Japan in accordance with Article 14, Paragraph 1 of the Japanese Basel Act. However, the exporter Nisso took no steps at all to implement these orders, and so the Government of Japan, as a substitute for Nisso, re-imported the cargo to Japan and disposed of it in accordance with the Act on Substitute Execution by Administration (Act No. 43 of 1948).

As mentioned above, the Japanese Basel Act applies *mutatis mutandis* to the import and export procedures stipulated by the Foreign Exchange Act. Consequently, if a "false statement" in the application for approval of export of the cargo was made, as in the Nisso incident, but the METI was not suspicious about the application, only the standard export procedure applies to the export under the Foreign Exchange Act, and thus the MOE would not start procedures based on the Japanese Basel Act independently. The Nisso incident made clear that the effectiveness of the export regulation stipulated by the Japanese Basel Act is not necessarily sufficient as a measure to prevent illegal traffic of hazardous wastes without the PIC procedure provided by the Basel Convention.

After the Nisso incident, the MOE and the METI took joint charge of enforcing the Japanese Basel Act and Japan Customs enforced the Act at the point of entry into and departure from Japan in an attempt to develop more effective measures for implementing the Basel Convention in Japan, especially fulfilling the PIC procedure stipulated by the Convention. Such efforts included (1) establishing the well-defined criteria for deciding whether or not the cargo to be imported or exported corresponds to the specified hazardous wastes under the Japanese Basel Act, (2) promoting the use of "Pre-Application Consultation Service" by importers, exporters and customs agents, etc., (3) sharing information obtained through the Pre-Application Consultation Service between the MOE, METI, and Japan Customs, and (4) holding training workshops for importers, exporters, and customs agents, etc. to provide information on not only the specified hazardous wastes under the Japanese Basel Act, but also recyclable resources, used items, and other material for the purpose of promoting their knowledge and understanding of the Japanese Basel Act, the Waste Management Act, the Foreign Exchange Act, the Customs Act, and other relevant laws and regulations.

Looking now at the well-defined criteria for deciding whether or not the cargo to be imported or exported corresponds to the specified hazardous wastes under the Japanese Basel Act, the MOE and the METI made public the criteria for containers made of polyethylene terephthalate (waste PET bottles, etc.), used tube televisions, and used products with built-in lead batteries.

The Pre-Application Consultation Service is intended to support importers and exporters in deciding whether or not items to be imported or exported correspond to the specified hazardous wastes under the

Japanese Basel Act and wastes under the Waste Management Act, and in taking necessary import/export procedures after receiving information concerning the relevant Japanese laws and regulations in advance of making application for import or export.

However, despite such efforts, many illegal traffic incidents have still occurred, in which exported cargos are recognized to be not "recyclable resources" or "used items" but only "wastes" by customs houses or quarantine stations in the importing States, and so are rejected and shipped back to Japan. Cargos that have been shipped back to Japan include (1) used lead batteries, which are designated as both hazardous wastes under the Basel Convention and the specified hazardous wastes under the Japanese Basel Act, (2) plastic wastes containing many other substances which are unable to be recycled because of their poor quality, and (3) televisions and computer monitors unapproved as used items because they were not able to be powered on during the inspection conducted at the customs houses and quarantine stations, etc. of the importing States.

The problematic situation still remains that, if Japan Customs fails to detect the making of "false statement" in the application for approval of export at the point of departure from Japan, the cargos containing the specified hazardous wastes under the Japanese Basel Act are exported from Japan to other States bypassing the export regulation procedures stipulated by the Act.

7.4   CONCLUDING REMARKS: MEASURES FOR IMPROVING THE ENFORCEMENT OF THE EXPORT REGULATIONS UNDER THE JAPANESE BASEL ACT

In light of the deliberation above, I would like to present some measures to improve the effectiveness of the export regulations under the Japanese Basel Act, from two points of view: (1) preventative measures to reduce the improper transboundary movements of hazardous wastes, and (2) ex-post measures after such movements have occurred.

Concerning preventative measures, the first issue to address is that attempts to export the specified hazardous wastes under the Japanese Basel Act are not punishable under current laws and regulations. When such an attempt is detected at the point of cargo inspection at customs before export, the exporter will be acquitted of the charge if he/she withdraws the application for export. As mentioned above, any person who has exported goods without obtaining approval from the METI shall be punished by "imprisonment with work for not more than 5 years or a fine of not more than 5,000,000 JPY, or both." However, the Japanese Basel Act criminalizes an attempt to export the specified hazardous wastes or a preparation to export such wastes, and so any person who attempts or prepares to export them shall not be punished at all.

In the Waste Management Act, the crime of an attempt to export waste and the crime of a preparation to do it were newly established by the 2005 amendment with the aim of increasing the deterrent effect on improper export of waste. One of the reasons for this amendment was the occurrence of an incident of improper but not necessarily illegal export of used plastic from Japan to Qintao, Shangdong Prefecture, in the People's Republic of China in 2004. Before amendment of the Waste Management Act in 2005, even if an improper export of waste was detected upon a pre-export cargo inspection at the Japan Customs, the exporter was not charged if the application for export was withdrawn, and so the Act's regulation of illegal export of waste was not necessarily sufficient in preventing it. The Act's 2005 amendment criminalizing both an attempt to export waste and a preparation to do it is intended to address this insufficiency.

Therefore, the Waste Management Act at present punishes not only the crime of having exported waste without obtaining confirmation from the MOE by "imprisonment with work for not more than 5 years or a fine of not more than 10 million yen" (paragraph 1(12), Article 25), but also the crime of an attempt to export waste without obtaining confirmation by "imprisonment with work for not more than 5 years or a fine of not more than 10 million yen" (paragraph 2, Article 25) and the crime of a preparation to do it by "imprisonment with work for not more than 2 years or a fine of not more than 2 million yen" (Article 27). Any representative of a juristic person shall also be punished by "a fine of not more than 300 million yen" for having exported waste or attempting to do it, and by "a fine of not more than 3 million yen" for preparing to export it (Article 32).

In order to improve the enforcement of the export regulations under the Japanese Basel Act, the establishment of crimes of an attempts to export the specified wastes and of a preparation to do so should be considered.

Second, concerning ex-post measures after improper transboundary movements of hazardous wastes, since the Nisso Incident occurred in 1999, the Government of Japan has taken only administrative measures against the exporters and others engaged in improper or illegal transboundary movements of hazardous wastes, such as "ordering them to explain the details" and "issuing a strict warning to them." However, cargo in cases of ship-back, in which case the cargo departing from Japan is not permitted to be unloaded at the port of the importing State and is shipped back to Japan, can be interpreted as "the exported cargo" under any of the existing relevant laws and regulations: the Japanese Basel Act, the Foreign Exchange Act, and the Customs Act. Therefore, if the exported cargo in a case of ship-back contains an item or a material regulated by these relevant laws and regulations and the export in question is considered to be malicious according to the amount or ratio of the regulated item or material, the Government of Japan should take stricter measures, including administrative sanctions to order the exporter to prohibit export, and judicial measures of criminal investigation, arrest, prosecution, and punishment of the exporter.

NOTES

1. The scope of application of the Basel Convention covers a wide range of wastes defined as "hazardous wastes" based on their origin, composition, and characteristics, as well as two types of wastes defined as "other wastes," including "wastes collected from households" and "residues arising from the incineration of households wastes."
2. However, after the adoption and enactment of the Basel Convention, African parties to the Basel Convention, supported by other developing parties and environmental NGOs, continued to insist on "a total world-wide ban," claiming that this was the only means to protect developing countries from becoming the "dumping grounds" of industrialized countries. The Third Meeting of the Conference of the Contracting Parties to the Basel Convention (COP 3), which took place in Switzerland in September 1995, adopted Decision III/1 (the so-called "Ban Amendment"), which amends the Convention to ban the movements of hazardous wastes destined for disposing of or recycling from the parties and other States listed in Annex VII (members of the OECD, EC, and Liechtenstein) to the parties and other States not listed in Annex VII. Nevertheless, entry into force of the Ban Amendment had been mired in controversy over the interpretation of paragraph 5 of Article 17 of the Basel Convention, which stipulates the number of ratifications, approval, formal confirmation, or acceptance by Parties needed to take amendments to the Convention into force, namely "at least three-fourths of the Parties who accepted them."

    The ground for breaking the deadlock was prepared by the "Country Led Initiative" (CLI). The President of the COP 9, which took place in Indonesia in June 2008, in his statement on the way forward on the Ban Amendment, called for a process to explore means by which objectives of the Ban Amendment might be achieved. Based on this statement, the Governments of Indonesia and Switzerland announced their readiness to organize the CLI, in order to discuss in an informal, dynamic, and non-dogmatic manner those issues related to the transboundary movements of hazardous wastes, especially to developing countries, contrary to the overarching objective of the Ban Amendment. The CLI process developed a draft omnibus decision, which formed the basis of discussions and a decision at the COP 10, which took place in Columbia in October 2010.

    Decision X/3 adopted at the COP 10 "Agrees, without prejudice to any other multilateral environmental agreement, that the meaning of paragraph 5 of Article 17 of the Basel Convention should be interpreted to mean that the acceptance of three-fourths of those parties that were parties at the time of the adoption of the amendment is required for the entry into force of such amendment, noting that such an interpretation of paragraph 5 of Article 17 does not compel any party to ratify the Ban Amendment." This decision clarifies the interpretation of paragraph 5, Article 17 of the Convention, setting the bar for entry into force of the Ban Amendment.

3. The other principles adopted by the Basel Convention are as follows: hazardous wastes moving across international boundaries must be in compliance with international rules and standards on packaging, labeling, and transportation. The Basel Convention also contains provisions regarding international cooperation and exchange of information.
4. Paragraph 1, Article 9 of the Basel Convention stipulates that: "For the purpose of this Convention, any transboundary movement of hazardous wastes or other wastes: (a) without notification pursuant to the provisions of this Convention to all States concerned; or (b) without the consent pursuant to the provisions of this Convention of a State concerned; or (c) with consent obtained from States concerned through falsification, misrepresentation or fraud; or (d) that does not conform in a material way with the documents; or (e) that results in deliberate disposal (e.g. dumping) of hazardous wastes or other wastes in contravention of this Convention and of general principles of international law, shall be deemed to be illegal traffic."
5. Paragraph 5, Article 9 of the Basel Convention stipulates that: "Each Party shall take appropriate legal, administrative and other measures to implement and enforce the provisions of this Convention, including measures to prevent and punish conduct in contravention of the Convention." Paragraph 5, Article 9 of the Convention stipulates that "Each Party shall introduce appropriate national/domestic legislation to prevent and punish illegal traffic."
6. Paragraph 3, Article 48 of the Foreign Exchange Act stipulates that: "In addition to cases prescribed in the preceding two paragraphs, the Minister of Economy, Trade and Industry may impose, pursuant to the provisions of Cabinet Order, on a person who intends to export specific kinds of goods or to export goods to the specified regions or a person who intends to export goods through specified transaction the obligation to obtain approval, to the extent necessary to maintain equilibrium of the international balance of trade, to achieve the sound development of foreign trade and the national economy, to sincerely fulfill obligations under the treaties and other international agreements Japan has signed, to make Japan's contribution to international efforts for achieving international peace, or to implement a cabinet decision set forth in Article 10, paragraph 1."

BIBLIOGRAPHY

Clapp, Jennifer (2001), *Toxic Exports: The Transfer of Hazardous Wastes from Rich to Poor Countries*, Ithaca: Cornell University Press.
Kofi Asante-Duah, D. and Imre V. Nagy (1998), *International Trade in Hazardous Waste*, London: E & FN Spon.
Krueger, Jonathan (1999), *International Trade and the Basel Convention*, London: Royal Institute of International Affairs.
Kummer, Katharina (1995), *International Management of Hazardous Wastes: The Basel Convention and Related Legal Rules*, New York: Oxford University Press.
Kummer, Katharina (1998), 'The Basel Convention: ten years on', *Review of European Community & International Environmental Law*, **7** (3), 227–36.
Louka, Elli (1994), *Overcoming National Barriers to International Waste Trade: A New Perspective on the Transnational Movements of Hazardous and Radioactive Wastes*, Dordrecht: Graham & Trotman.
Ministry of Environment, the Government of Japan. Suishitsu-hozen-kyoku, Haikibutsu-mondai-kenkyukai (Water Quality Bureau, Hazardous Waste Study Group) (ed.) (1993), *Bazeru Shinpou Q and A* (in Japanese) (The Japanese Basel Act Questions and Answers), Tokyo: Daiichi-Houki.
Ministry of Environment, the Government of Japan (2009), *Japan's Activities*

for Implementing the Basel Convention, Tokyo: Ministry of Environment, Government of Japan.

Moyers, Bill and the Center for Investigative Reporting (1990), *Global Dumping Ground: The International Traffic in Hazardous Waste*, Washington: Seven Locks Press.

Tsuruta, Jun (2005), 'Kokusai Kankyo Jouyaku ni okeru Jouyaku Jitsussen no Doutaiteki Katei' (in Japanese) (Dynamic Treaty Process of International Environmental Framework Convention), in Shiroyama Hideaki and Yamamoto Ryuji (eds), *Kankyo to Seimei* (Environment and Life), Tokyo: University of Tokyo Press.

Tsuruta, Jun (2007), 'Kokusai Shigen Junkan no Genjou to Kadai' (in Japanese) (The Present Status and its Problem of International Resources Recycling), *Hogaku Kyousitsu*, **326**, 6–12.

Tsuruta, Jun (2012), 'Basel Jouyaku to Basel Hou' (in Japanese) (The Basel Convention and the Japanese Basel Act), in Niimi Ikufumi, Matsumura Yumihiko and Otsuka Tadashi (eds), *Kankyou Hou Taikei* (Outline of Environmental Law), Tokyo: Shouji-Houmu.

APPENDIX 7.1

Article 1:   Purpose

The purpose of this Act is to take measures to regulate the export, import, transportation and disposal of the specified hazardous wastes, etc. in order to ensure accurate and smooth implementation of "the Basel Convention on the Control of Transboundary Movements of Hazardous Wastes and Their Disposal" (hereinafter referred to as "the Convention") and other agreements, and thereby to contribute to the protection of human health and the sound living environment.

APPENDIX 7.2

Article 2:   Definition of the Specified Hazardous Wastes

Paragraph 1

(1)   In this Act, the "specified hazardous wastes, etc." means the following materials (excluding the wastes generated in association with the navigations of vessels that are specified by Government Ordinance, and radioactive materials and the materials contaminated by such radioactive materials):

(i)   Materials to be exported or imported for the disposal operations contained in Annex IV of the Convention (hereinafter referred to as "disposal") that fall under any of the following:

(a)   Materials contained in Annex I of the Convention that have any of the hazardous characteristics contained in Annex III of the Convention

(b)   Materials contained in Annex II of the Convention

(c)   Materials on which the Government of Japan, as provided by Government Ordinance, has notified the Secretariat of the Convention in accordance with Paragraph 1 or 2, Article 3 of the Convention

(d)   Materials for which notification was received from the Secretariat of the Convention in accordance with Paragraph 3, Article 3 of the Convention, that are specified by the Ministry of the Environment Ordinance as materials pertaining to export, with the region pertaining to the notification in question as the region of destination or transit, or as materials pertaining to import, with such region as the region of origin, shipment or transit.

(ii) Materials for which the export, import, transportation (including storage in association with these; hereinafter the same shall apply) and disposal need to be controlled based on bilateral, multilateral or regional agreements or arrangements specified in Article 11 of the Convention (hereinafter referred to as "agreements, etc. other than the Convention") that are specified by Government Ordinance.

APPENDIX 7.3

Article 4:   Export Approval

Paragraph 1
Any person who intends to export the specified hazardous wastes, etc. shall be obliged to obtain an export approval in accordance with the provision of Article 48, Paragraph (3) of the Foreign Exchange and Foreign Trade Law (Act No. 228 of 1949).

Paragraph 2
When an application for the export approval set forth in the preceding paragraph is made for the specified hazardous wastes, etc. that are destined for the region specified by the Ministry of Economy, Trade and Industry and the Ministry of the Environment joint ordinance where it is particularly necessary to prevent air pollution, water contamination and other environmental pollution (hereinafter simply referred to as "environmental pollution") that is likely to occur in association with the disposal of the specified hazardous wastes, etc. to be exported, the Minister of Economy, Trade and Industry shall send a copy of the written application to the Minister of the Environment.

Paragraph 3
The Minister of the Environment shall, when the written application is sent to the Minister in accordance with the provision of the preceding paragraph, confirm whether necessary measures to prevent environmental pollution are taken in regard to the disposal of the specified hazardous wastes, etc. pertaining to the written application, and notify the Minister of Economy, Trade and Industry of the confirmation result.

Paragraph 4

The Minister of Economy, Trade and Industry shall not approve the export set forth in Paragraph 1, until the Minister receives notification from the Minister of the Environment that necessary measures to prevent environmental pollution are taken in accordance with the provision of the preceding paragraph.

8. Impact of the raw material import duty reduction system on international waste trading

So Sasaki

INTRODUCTION

Many Asian countries have pursued export-oriented industrialization for their growth strategies. One of the core policy measures for export-oriented industrialization is the establishment of export processing zones (EPZs) using the raw material import duty reduction system (Table 8.1). Amano (2005) pointed out that "in Asian countries, although the contents of export-oriented industrialization policies are characterized by individual countries and regions, the policies oblige foreign companies to export their products, in return for providing the import permit and the duty exemption on materials and/or equipment, and the other tax incentives." Many foreign companies have invested in export-oriented industrial parks in developing Asian countries such as China, Thailand, Malaysia, Indonesia, and the Philippines. They import raw material for the processing and assembly of their products, and export the products to other countries. Import duty reduction systems stimulate investment incentives for foreign companies.

However, the import duty reduction system can be utilized as a route for smuggling into the domestic market. Many Asian countries are eager to protect their market from the inflow of cheaper foreign goods on which taxes are not levied. As a result, the shipment of raw materials, parts, and products from EPZs to other domestic markets is often strictly controlled.

The shipment of waste from EPZs to domestic waste treatment and recycling companies is also strictly controlled and managed. If a factory located in an EPZ intends to send its waste to domestic recycling facilities outside of the EPZ, the waste generator must submit certain documents to the authorities and pay an import tariff. It is also typical for the authority or management company operating the EPZ to specify a certain waste collection company providing services to the zone.

Table 8.1   Material import duty reduction systems in Asian countries

Country	Related Law
Indonesia	No.111/PMK.010/2006
	No.129/KMK.04/2003
Thailand	Refund of parts and material import duties for export by clause 2 of customs law of Article 19
	Reduction of tariff to goods kept based on bond scheme
	Exemption from taxation on articles brought in to free zone (FZ)
	Exemption from taxation on articles brought in to export processing zone (EPZ)
	Exemption from taxation on articles and material import duties for exports based on the Board of Investment scheme
China	Tariff measures on processing trade
Philippines	Customs Administrative Order-CAO, No.12-2003
Vietnam	Circular 04/2007/TT-BTM
	Official Letter 4537/TCHQ-KTTT
Malaysia	Exemption from customs duty to product put on Free Zone (FZ)

Source: JETRO website.

As a result of these restrictions, factories located in EPZs have often sent their waste to recyclers in other countries. Recyclable waste that can be treated in another country is also exported. Many developing Asian countries are net importers of metal scrap and waste paper; however, a portion of the exported recyclable waste such as metal scrap and waste paper from EPZs can be viewed as unnecessary movement.

On the other hand, some waste should be exported to other countries if the appropriate technologies to manage it are lacking in the country where it is generated. If the destination of the waste has environmentally sound facilities with high efficiency material recovery, the export of waste is beneficial. While several studies have looked at waste recycling, treatment infrastructure, and policies with regard to international waste trading, little attention has been paid to the raw material import duty reduction system for export industries, which has the unintentional effect of increasing international waste trading. Up until now, international waste trading and import duty reduction systems have been discussed separately, and have not been examined in the English literature thus far. In a Japanese article, Sasaki (2007) examined the case of Thailand where international waste trading was encouraged because of a raw material import duty reduction system that acted as a block on the development of recycling industries in developing Asian countries. This chapter focuses on the recycling of waste generated in EPZs. The first section investigates the impact of this system on international waste trading with China, Thailand, the Philippines, Indonesia, Vietnam, and Malaysia. The second section focuses on the effect of Japanese companies in developing Asian countries on the import duty reduction system and the recycling market in Asia. The third section deals with the impact of free trade agreements (FTAs) and economic partnership agreements (EPAs) on EPZs and waste flow. Policy implications are discussed in the fourth section.

8.1   CASE STUDIES OF THE IMPACT OF RAW MATERIAL IMPORT DUTY REDUCTION SYSTEMS ON THE PROMOTION OF INTERNATIONAL WASTE TRADING

This section discusses case studies of export industries in China and Thailand, which are using a raw material import duty reduction system, and how these industries recycle and treat materials and parts.

8.1.1   The Case of China

E & E Solutions (2005, pp. 106–7) report that Japanese export industries must adhere to the following rules and regulations in China if they want to send waste to local companies. Export industries must pay a "value added tax" and follow procedures such as weight measurement and evidence photography if they sell scrap used in a raw material import duty reduction system for Chinese domestic recycling companies. If they sell scrap from a free trade zone (FTZ) to a domestic recycling company outside the FTZ, they must pay a "value added tax" and an "export tax." Therefore, exporting to Japan can be a way to avoid the former tax. Companies export material as scrap, send it back as defective products, or sell it to Japan as material rather than waste. If they export it, they do not have to pay tax. Moreover, export costs tend to be cheaper than recycling costs in China.

If Japanese companies entrust waste disposal to domestic companies, they must use companies assigned by the environmental protection department and customs department of the province. Further, they cannot sell recyclables and must pay a disposal fee. Japanese companies that do not want to pay taxes sometimes export waste as products to Hong Kong from China. Hong Kong's traders then re-export the waste to China. The reason for such trade is that China and Hong Kong signed the Closer Economic Partnership Agreement (CEPA) in 2004, eliminating import and export tariffs between the two (Shinko Research Co. Ltd 2008, p. 136). The FTZ system in China creates an unnecessary burden on export industries for waste disposal and recycling.

8.1.2 The Case of Thailand

In Thailand, a raw material import duty reduction system has been adopted by the Board of Investment (BOI), and many export industries use this system. Recycling and waste processing is conducted in the following manner.

When export industries sell defective goods and process waste to a domestic recycling company, they must pay a "value added tax" and an "export tax" that is based on the sales price. If they use local companies to collect and recycle, they must ask a BOI officer to inspect. Inspection takes 30 to 40 days and costs 5,000 Baht (around 163 USD) each time. Inevitably, export industries select recycling companies located in EPZs or duty free zones (DFZs). Export industries receive two advantages: avoiding inconvenient procedures as well as taxes.

If recycling companies based in an EPZ or DFZ accept hazardous waste, they must export to Japan based on Basel Convention regulations, but may export to Hong Kong or Singapore without following them. In the latter cases, the waste is recognized as non-hazardous waste in Hong Kong or Singapore. Therefore, recycling companies prefer to export to Hong Kong and Singapore, which have less strict customs inspection compared with Japan and other free trade ports. On the other hand, domestic recycling companies not located in a FTZ or DFZ have a tax issue to receive waste from EPZ or DFZ. According to UNI Copper Trade Ltd, an electronic waste recycling company, export industries must pay not only a value added tax and export tax, but also may be charged a fee of four times the initial tax if they fail to pay and customs officials discover this (JETRO 2004, pp. 2–61). Thus, UNI Copper Trade Ltd established a related company in a FTZ in 2007 to expand the collection amount.

8.1.3 The Case of Other Asian Countries

In the Philippines, if export industries located in a FTZ approved by the Philippine Economic Zone Authority (PEZA) use local companies to recycle their scrap, they are charged a tax. Therefore, Japanese recycling companies based in the country have two factories inside and outside the FTZ to collect scrap. Philippine Business for the Environment, a non-governmental organization, negotiates with the PEZA for local recycling companies to collect scrap, and schedules events several times per year to collect scrap without duty from FTZs.

In Indonesia, if export industries sell defective goods and process waste to domestic recycling companies, they must pay an export tax and ask a customs officer to inspect.

In Vietnam, export industries also use recycling companies which export to other countries when parts and raw materials from a raw material import duty reduction system become process waste.

In Malaysia, export industries located in the free zone (FZ) are exempt from import taxes on the condition that they export more than 80 percent of their products. Thus, when export industries sell defective goods and process waste to a domestic recycling company, they must pay a tax. On the other hand, recycling companies are charged a 10 percent export tax if they export scrap. Consequently, recycling companies located in the FZ do not benefit from collecting scrap from domestic companies. So it seems reasonable to conclude that the domestic and FZ recycling markets in Malaysia are more separated than in other Asian countries.

Cambodia has EPZs; however, unlike other Asian countries, there are no laws precluding domestic recycling with use of a material import duty reduction system on the grounds that the procedure becomes complicated and disadvantageous in attracting foreign investment.

8.1.4 Summary

As mentioned above, we can conclude that a raw material import duty reduction system has the effect of increasing international waste trading because export industries try to avoid inconvenient procedures and taxes, making it difficult for local recycling companies to collect scrap and recyclable materials. Consequently, this can be considered as one factor contributing to the importation of hazardous waste to Japan from other Asian countries.

However, it is not easy to analyze the quantitative impacts of raw material import duty reduction systems on the promotion of international waste trading because raw materials in a duty reduction system are converted to products (and HS Code) and are exported as semi-products or sundries.

8.2   RESPONSES BY JAPANESE COMPANIES

Some portion of waste scrap from export industries is generated from stamping, off-specs, and inferior goods, which comprise about 10 percent to 20 percent of the assembled parts on a product weight basis during final assembly. If parts and raw materials used in a raw material import duty reduction system become process waste, it is necessary for companies to know whether this waste is processed or recycled. If it is sold as parts or materials in the domestic market, it is regarded as violation of the import duty reduction system. In addition, if defective parts or inferior goods sold in the market caused problems or accidents, the brand image and credibility of the manufacturer could be damaged.

To avoid such situation, factories located in EPZs have two options. The first is to export their waste to other countries; the second is to send their waste to reliable recycling companies, which are able to show traceability of waste flow and treatment to the waste generator. This section focuses on Japanese factories' response to an import duty reduction system. Section 8.2.1 focuses on the transboundary movement of hazardous waste from Asian countries to Japan, which is a destination of waste generated in export-oriented zones. Sections 8.2.2 and 8.2.3 present two case studies of Japanese recycling companies located in China and Thailand, respectively. These firms try to recycle waste from foreign companies located in China and Thailand.

8.2.1 Transboundary Movement of Hazardous Waste from Asian Countries to Japan

Hazardous wastes are imported to Japan from Asian countries every year. According to Japan's Ministry of the Environment, the amount of hazardous waste imported between 1999 and 2008 was 41,291 tons, of which 95 percent was imported from Asian countries.[1] Almost all hazardous waste received in Japan was exported by Japanese companies based in Asian countries; rather than treating the hazardous waste in the country where waste was generated, Japanese enterprises exported it to Japan for treatment.

In 2010, 4,292 tons of hazardous waste was imported from all Asian countries. By country, the largest exporter was Thailand, at 1,233 tons, followed by Singapore with 831 tons. The largest single item amount was electronics scrap from Thailand at 962 tons. The second largest single item amount was also electronics scrap, from Hong Kong, but it should be pointed out that the majority of this scrap originated in mainland China (Figure 8.1).

Why is hazardous waste being imported to Japan? What problems do Japanese enterprises in Asian countries face in regard to conducting hazardous waste treatment in the country where waste was generated?

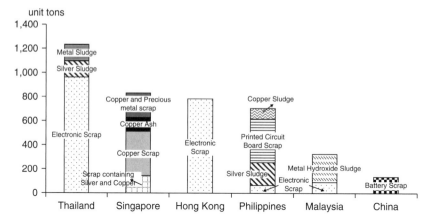

Note: Items under 100 tons are omitted.

Source: Ministry of Environment, Japan (2011).

Figure 8.1   Amount of hazardous waste imported based on the Basel Convention in 2010

Asian governments have taken measures to invest in waste recycling and treatment infrastructure. In fact, waste recycling and treatment industries not only in Japan, but also throughout Asia, have continued to develop, although some recycling processes for specific materials do not exist in each country. A questionnaire survey of Japanese enterprises in Asian countries in 2002 showed that it was difficult to recycle and to treat waste such as electronics scrap and sludge (see Figure 8.1).[2] However, we cannot conclude that there is a shortage of waste recycling and treatment facilities from these results alone. According to the above mentioned study, when Japanese enterprises operating in Asia were asked about problems with waste disposal (Figure 8.2), "lack of information about the disposal company" was the most common response at 39.5 percent, followed by "after waste was sent to the waste treatment and disposal company, it is not clear whether the waste has been properly processed" at 38.7 percent. Only 20.6 percent of respondents answered that "there is no proper processing company."

In regard to problems with waste recycling (Figure 8.3), "lack of information and technology for recycling" was the most common response at 42.2 percent, followed by "cannot effectively recycle due to small amount of waste," "recycling cost is too high," and "after the waste was sent to a recycling company, it is not clear whether the waste has been properly recycled or not," each reported by 27.3 percent of respondents. "There is no proper recycling company" was reported by 25.5 percent. This study

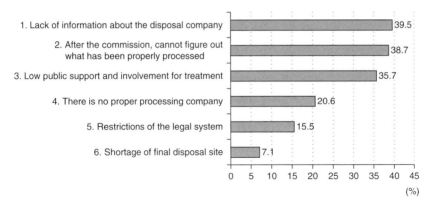

Source: Kyushu Bureau of Economy, Trade, and Industry (2003).

Figure 8.2 *Problems of waste disposal for Japanese enterprises in Asian countries*

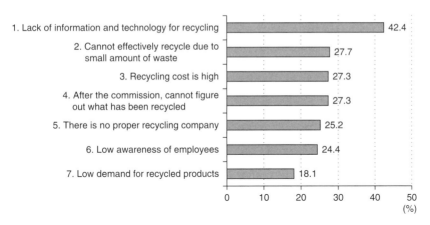

Source: Kyushu Bureau of Economy, Trade, and Industry (2003).

Figure 8.3 *Problems of waste recycling for Japanese enterprises in Asian countries*

was conducted in 2002, when the recycling industry in developing Asian countries was less developed than it is now. These findings show a high level of concern over the information available about recycling technology and ensuring proper treatment.

However, Asian recycling and treatment industries, including Japanese recyclers in other Asian countries, have clearly developed since then. Sasaki (2008) and Mitsubishi UFJ Research and Consulting (2009a) show that it is possible in the Asian region to recycle and to treat the waste shown in Figure 8.1 and inquired about in the questionnaire. Yet, there remain not only problems of physical infrastructure such as a shortage of recycling and treatment facilities, but also systematic problems related to issues such as law enforcement, recycling costs, level of technology, and assurance of proper disposal and recycling.

As shown in Section 8.1, import duty reduction systems are a driving force of hazardous waste exports to Japan. Many recycling companies in Japan have been trusted by Japanese manufacturers because they have good business relationships. Some recycling companies also have invested in other Asian countries, mainly to treat recyclable waste from Japanese manufacturers located in the country.

8.2.2   Case Study of Recycling Company D in China

Recycling company D is owned by a major Japanese nonferrous smelting company and was founded in 2003. They recover precious metal and treat hazardous waste from electronics scrap, metal plating waste, and other materials that are generated by foreign-invested and especially Japanese companies across Jiangsu province in China. The treatment process is the same as that in Japan and utilizes pyrometallurgical and hydrometallurgical processing.

Shimada (2005) says that recycling processes in China need advanced integration of pre-treatment, copper smelting, lead smelting, zinc smelting, and precious metal refining technologies. In the course of doing business in China, it is necessary to build trust with local companies with regard to sampling and analysis of scrap with precious metal content. Shimada also points out that it is difficult to ensure traceability of waste as it is often sold into the second-hand market or exported. It is often difficult to track factory scrap beyond the province, because of the vertical division of local institutions on hazardous waste and transportation. Further, local companies utilizing pyrometallurgy have no concerns about air pollution and do not use secondary combustion furnaces to decompose toxic gases after recovering gold. The weak enforcement of pollution control regulation promotes such dirty recycling.

As a result, company D is at a disadvantage compared to local companies, because local companies can reduce pollution control costs so that they can offer a higher price for waste than company D to compete for the scarce resources (JETRO Beijing Center 2009, pp. 79–81).

8.2.3   Case Study of Recycling Company H in Thailand

Japanese recycling company H established their subsidiary in Thailand in 2003 to recycle scrap generated from small and medium-sized Japanese electronics manufacturers in the country. The two reasons for focusing on small and medium-sized customers are as follows. First, it is more difficult for them to manage industrial waste than larger companies because of their lack of human resources and small levels of waste generation. Second, the stringent quality requirements of Japanese manufacturers, who are their main customers, lead to poorer waste yields.

Recycling company H built up their own recycling network linking Japan and Thailand in order to improve their recycling rate after separation and dismantling. Utilizing their network, they increase the number of items purchased from customers, reduce processing costs, and provide low service prices. They cannot treat all of the waste in Thailand, so some is returned to Japan to ensure the recycling flow shown in the original manifests for guaranteed traceability.

However, in 2006 the subsidiary was sold by the Japanese headquarters to a local company following decreased profitability. Soaring commodity prices had led to intensive price competition and increased cost of collecting electronic scrap. This led to increased difficulty in the collection of electronic scrap and lower profits.

8.2.4   Summary

In both cases, the Japanese recycling companies had a high level of recycling technology and services to ensure proper recycling. However, competition with local companies led to intense price competition that hampered their ability to collect waste and recyclable materials. In other words, the current situation of the recycling market in Asia is that "bad money drives out good." The problems of systematic infrastructure go deep into the heart of the recycling market in Asia.

Those Japanese companies that recycle hazardous waste by exporting to Japan have lower price competitiveness than Japanese recycling companies that use the local recycling market in Asia. This is because the international trade of hazardous wastes requires additional transaction costs related to prior notice and consent procedures in order to comply with the Basel Convention as well as the customs regulations.

Therefore, although systematic infrastructure problems are indeed key factors in the importation of hazardous waste into Japan, it is difficult to explain such importation with these reasons alone. Thus, the raw material import duty reduction systems that have been established in Asian countries impact the international trade of recyclable waste.

8.3   IMPACT OF FTA AND EPA

It was pointed out in Section 8.1 that export industries prefer to export their waste to Hong Kong and Singapore because of less strict customs inspection compared with Japan and free trade ports. This section discusses the impacts of import taxes, as well as FTAs and EPAs, on the current international flow of recyclable waste generated in EPZs.

With regard to import taxes, it is easier to export processed waste under a raw material import duty reduction system to Hong Kong and Singapore because they are free trade ports. However, according to Japan's Tariff Schedule (January 2012), most recyclable materials are tax-free except for items such as rare metal scraps (Table 8.2). As a result, we cannot conclude that the level of tariff determines the destination of the processed waste. It has been pointed out that Japan was late in concluding

Table 8.2   Japanese tariff rates and amount of recyclable materials imported in 2010

HS-Code	Description	Tariff rate (%)	Amount of import (kg)
7204.50010	Remelting scrap ingots of alloy steel	5.7	1,140
7204.50020	Other remelting scrap ingots of alloy steel	4.1	129
8112.22000	Waste and scrap of chromium	4.1	27,080
8112.29000	Other Waste and scrap of chromium	5.2	88,634
8112.52000	Waste and scrap of thallium	4.1	0
8112.59000	Others of beryllium, chromium, germanium, vanadium, gallium, hafnium, indium, niobium (columbium), rhenium and thallium, and articles of these metals, including waste and scrap	5.2	7
8112.92100	Unwrought waste and scrap, powders of vanadium	3.0	417,068
8112.92200	Unwrought waste and scrap, powders of indium	5.2	70,100
8113.00000	Cermets and articles thereof, including waste and scrap	5.2	149,426

Source: Customs and Tariff Bureau, Ministry of Finance (2012).

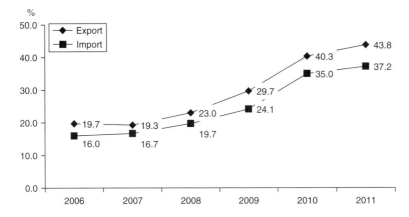

Source: JETRO (2011).

Figure 8.4   Rate of use of FTAs and EPAs by Japanese industries in ASEAN Countries

FTAs or EPAs compared to other countries, but it seems unlikely that this is the reason that export industries prefer to export to Hong Kong and Singapore, as JETRO analysis has shown that Japanese export industries in Asia certainly have increased the use of the FTAs or EPAs year on year (JETRO 2011, Figure 8.4). Additionally, according to a recent questionnaire survey conducted by JETRO, when asked the reasons for not using an FTA or EPA, respondents answering that they "already have advantages without the FTA and EPA with regard to import taxes" decreased from 35.4 percent to 14.5 percent (JETRO 2009, 2011). If this were indeed the case, it would be better for Japanese export industries to select domestic recycling, rather than export to Singapore, Hong Kong, or Japan. However, even if FTAs and EPAs are implemented further, it cannot be confirmed from the field surveys of recycling companies or in the statistics related to waste management that the structural impacts of a raw material import duty reduction system on increasing international waste trading would change. There is clearly room for considering some other obstacles to sending waste to domestic recycling companies.

8.4   POLICY IMPLICATIONS

As discussed in Section 8.2, in developing Asian countries, recycling companies with environmentally sound technology and management face difficulty in collecting waste because of competition with recyclers without proper pollution control. Import duty reduction systems, which become obstacles to sending processed waste to domestic recyclers, hinder waste destined for domestic recyclers without pollution control. However, they also discourage investment in environmentally sound technology in the country. To promote domestic recycling industries, revision of import duty reduction systems should be considered. From the viewpoint of the roles of EPZs and considering environmentally sound recycling, the revision should include traceability systems, which assure the destination of waste, in addition to import duty exemptions or lower tariff rates.

Some recyclable waste should be exported, however, if environmentally sound recycling facilities are not located in a country. The field surveys of recycling companies revealed that a procedure based on the Basel Convention is needed when the waste is exported to Japan, but is not needed when it is exported to Hong Kong and Singapore. If the wastes are defective printed circuit boards, it seems reasonable to conclude that Hong Kong and Singapore have less stringent customs inspections than Japan.[3]

Regarding printed circuit board waste and waste containing nonferrous metals, Japanese recycling industries have an advantage in the high-efficiency recovery of various metals. With regard to the recycling of rare metals, technologies combining pyrometallurgical and hydrometallurgical recycling used by Japanese nonferrous smelting companies are more efficient and environmentally friendly than local Asian hydrometallurgical recycling. Japanese technology for nonferrous smelting is essential for recycling rare metals that are not now recycled.[4] It seems reasonable, therefore, to conclude that strategic Japanese policies such as a deregulation of domestic rules pertaining to the Basel Convention are required in order to expand the amount of recyclable materials imported for nonferrous smelting.

A further direction of the work outlined in this chapter will be to provide more evidence relating to environmental impact assessments and material flow analysis. More research on international waste trading in both the manufacturing and recycling industries is necessary.[5]

8.5   CONCLUSION

This chapter showed from case studies in Asian countries that it is important to recognize that the problems related to the recycling and treatment industries in Asia are not only problems of physical infrastructure such as a shortage of recycling and treatment facilities, but also problems of systematic infrastructure such as law enforcement, recycling costs, the level of technology, and guarantee of proper disposal. We also showed the structural impacts of raw material import duty reduction systems on promoting international waste trading.

It is concluded that a raw material import duty reduction system as a part of an export-oriented industrialization policy as adopted by some Asian countries has the effect of increasing international waste trading. Export industries do their best to avoid inconvenient procedures and taxes, making it difficult for local recycling companies to collect scrap and recyclable materials.

What needs to be emphasized is that it is possible to re-examine import duty reduction systems to facilitate domestic recycling and provide assurances on the destination of waste. Regarding printed circuit board and nonferrous metal wastes, Japan should deregulate domestic rules for the Basel Convention to expand the amount of recyclable materials imported for the nonferrous industry, which is a process in which Japan has an advantage in recovering various metals, including rare metals.

NOTES

1. The amount of traded hazardous waste is based on permits issued by the Ministry of Environment, Japan.
2. Kyushu Bureau of Economy, Trade, and Industry (2003). The number of respondents is 779 companies. In these survey results, waste oil, waste plastic, glass cullet, and containers polluted by chemicals are excluded from the definition of hazardous waste imported to Japan.
3. It is less probable that officials in Hong Kong and Singapore admit that the Basel Convention is loosely enforced in their regions. However, imports to Singapore and Hong Kong of e-waste not approved by the Basel Convention are reported by Kojima (2005, pp. 125–7) and Mitsubishi UFJ Research and Consulting (2009b, pp. 76–92).
4. According to field surveys of Japanese nonferrous smelting companies, the yield of local recycling companies in Asia is about 80 percent, and local recycling companies in Asia sometimes request Japanese nonferrous smelting companies to recycle 20 percent of the residue.
5. For example, Shimizu and Sasaki (2009) examine the material flow in each country of Asia of the rare earth permanent magnets, and consider how recycling contributes to the sustainable supply of the raw material in the entire supply chain.

REFERENCES

Amano, T. (2005), *Higashi Asia no Kokusai Bungyou to Nihon Kigyou* (The International Division of Labor and Japanese Industries in East Asia), Yuhikaku.

E & E Solutions Co. Ltd (2005), *Nichu Kokusai Shigen Junkan Jittai Chousa* (International Recycling Survey between China and Japan), Ministry of Economy, Trade and Industry.

Japan External Trade Organization (JETRO) (2004), *Thai Koku Ni Okeru Kaden*

Seihin Tou No Haiki Oyobi Recycle No Genjyou Ni Kansuru Chousa (The Current Situation of Disposal and Recycling of Home Electric Appliances in Thailand).

JETRO (2009), *Zai ASEAN 7kakoku, India, Oceania, Shinsyutsu Nikkei Kigyou No FTA/EPA No Siyou Jyoukyou Ni Kansuru Chousa Kekka Gaiyou* (Summary of the Actual Use FTA/EPA by Japanese Industries in ASEAN 7 Countries, India, and Oceania).

JETRO (2011), *Zai Asia/ASEAN Nikkei Kigyou Katsudou Jittai Chousa* (The Current Situation of Japanese Industries Activities in ASEAN).

JETRO Beijing Center (2009), *Chougoku No Kankyou Sangyou Ni Kansuru Chousa Houkokusyo* (The Survey of Environmental Industries in China).

Kyushu Bureau of Economy, Trade and Industry (2003), *Asia Shinsyutsu Nikkei Kigyou tou Shigen Junkan Taiou Needs Chousa* (Needs Assessment of Recycling from Japanese Industries in Asia), Kyushu Bureau of Economy, Trade and Industry.

Kojima, M. (2005), 'Tounan Asia Shokoku Ni Okeru Junkan Shigen No Ekkyou Idou' (Transboundary Movements of Recyclables in Southeast Asia), in M. Kojima (ed.), *Asia Ni Okeru Junkan Shigen Boueki* (Trade of Recyclables in Asia), Institute of Developing Economies, pp. 117–32.

Ministry of Economy, Trade and Industry (METI) (2009), *Rare Metals Kakuho Senryaku* (Strategy of Securing Rare Metals).

Ministry of Finance (2012), 'Japan's Tariff Schedule as of January 1 2012', http://www.customs.go.jp/english/tariff/2012_1/index.htm (accessed 14 January 2012)

Mitsubishi UFJ Research and Consulting Co. Ltd (2009a), *Asia 3R System Koutiku Kiso Chousa Jigyou Houkokusyo* (The Basic Research of Establishment of 3R System), Ministry of Economy, Trade and Industry.

Mitsubishi UFJ Research and Consulting Co. Ltd (2009b), *Basel Houkisei Taisyou Kijyun Sakusei Tou Chousa* (Survey of Creating Standards for Basel Convention), Ministry of Environment.

Sasaki, S. (2007), 'Tokushu Asia Ni Okeru 3R-Thai Ni Okeru Seizougyou Ni Taisuru Zeisei Yugu Seido to Recycle' (Special edition, 3R in Asia, The Recycling and Impact of Raw Material Import Duty Reduction System in Thailand), in Asiken World Trend, Institute of Developing Economies, pp. 20–23.

Sasaki, S. (2008), 'Thai Ni Okeru Sangyou Haikibutsu Shori to Recycle No Genjyou – Kisei Kanwa Seisaku Wo Chushi Ni Shite-' (The Current Situation of Industrial Waste Management and Recycling in Thailand), in M. Kojima (ed.), *Asia Ni Okeru Recycle* (Recycling in Asia), Institute of Developing Economies, pp. 193–224.

Shimada, K. (2005), 'Kikinzoku Recycle To Kanren Gijyutsu No Fukyu Iten' (Metal Recycling and the Spread and the Transfer of the Related Environmental Technology), http://www.env.go.jp/council/06earth/y065-05/mat02.pdf (accessed 14 January 2012)

Shimizu, K., and S. Sasaki (2009), 'Kidorui Sangyou No Kodoka Ni Muketa Recycle –NdFeB Jisyaku Wo Rei Ni-' (The Recycling for Advancement of Rare Earth Industries, Case studies NdFeB magnet), in *Nihon Kidorui Gakaki Shi* (Journal of The Rare Earth Society of Japan), **55**, 49–57.

Shinko Research Co. Ltd (2008), *Koubutsu Shigen Kyoukyu Taisaku Chousa Houkokusyo* (Survey of Mineral Resources Supply Measures), Agency for Natural Resources and Energy.

9. Lessons learned from illegal transboundary movement of hazardous waste in Asia

Michikazu Kojima, Aya Yoshida, So Sasaki, and Sungwoo Chung

INTRODUCTION

Various forms of hazardous waste have been exported from developed countries to developing countries, causing environmental pollution. In the 1980s, hazardous waste imported into developing countries was often dumped improperly and resulted in health hazards. Recycling of hazardous waste, such as lead-acid batteries, has also caused pollution in importing countries. To solve these problems, the Basel Convention was established in 1989 and came into effect in 1992 after dozens of countries ratified it. Accordingly, prior notice and consent is required to export hazardous waste to other countries before shipment commences. As of April 2012, the number of ratifying parties had reached 178.

Member countries have introduced domestic regulations in line with the Basel Convention. In addition, some countries have started regulating the international trade of non-hazardous recyclable waste to prevent pollution problems from the process of recycling (see Chapters 1, 3, 4, and 5, and Kojima (ed.) 2005). Even after the Basel Convention and related regulations were implemented, many illegal shipments were intercepted by officials in importing countries, some of which were shipped back to their country of origin.

Serious health hazards caused by imported hazardous waste in developing countries have led to calls for a ban on hazardous waste exports to developing countries. For example, in August 2006, improper dumping of hazardous waste unloaded in Cote d'Ivoire from the *Probo Koala*, a vessel coming from Europe, caused the deaths of ten people and hospitalized 69.[1] Three months after the incident, the 8th Conference of Parties of the Basel Convention (COP) was held in Nairobi. At the COP, many developing countries, especially those in Africa, requested an early entry into force of the Ban Amendment which prohibits the export of hazardous waste from annex VII countries (developed countries) to non-annex VII countries (developing countries). However, the waste unloaded to Cote d'Ivoire was not under the jurisdiction of the Basel Convention, because the waste was generated from washing out tanks on the ship. In fact such waste generated on ships is under the jurisdiction of the International Convention for the Prevention of Pollution from Ships (MARPOL), which was adopted in 1973. In this case then, the Ban Amendment cannot be regarded as an effective measure for preventing illegal dumping of hazardous waste generated on ships.

It is important, therefore, to analyze cases of illegal shipment of hazardous waste in order to identify effective and appropriate preventive measures and to treat such waste in an appropriate manner. While some cases have been reported in newspaper articles and official documents, no comprehensive analysis has been conducted. In this chapter, we review a number of uncovered cases of illegal transboundary shipments of hazardous and recyclable waste in Asia, and extract lessons from the findings. In Section 9.1, the legal text and Guidance Document of the Basel Convention (Secretariat of the Basel Convention 2002) related to illegal shipment is explained. In Section 9.2, some cases of illegal transboundary movement of waste are reviewed, and in Section 9.3, key lessons and points of comparison are discussed. Section 9.4 extracts lessons from the case studies presented and argues for countermeasures to prevent and address illegal shipment.

9.1   ILLEGAL TRANSBOUNDARY MOVEMENT OF HAZARDOUS WASTE AND RECYCLABLE WASTE

9.1.1   Illegal Traffic Definition in the Basel Convention

Illegal transboundary movement of hazardous waste is defined in Article 9 of the Basel Convention as

> any transboundary movement of hazardous wastes or other wastes: (a) without notification pursuant to the provisions of this Convention to all States concerned; or (b) without the consent pursuant to the provisions of this Convention of a State concerned; or (c) with consent obtained from States concerned through falsification, misrepresentation or fraud; or (d) that does not conform in a material way with the documents; or (e) that results in deliberate disposal (e.g. dumping) of hazardous wastes or other wastes in contravention of this Convention and of general principles of international law.

When an illegal transboundary movement is uncovered, the responsible stakeholders should be identified. If the illegal traffic is the result of conduct on the part of the exporter or generator, "the State of export shall ensure that the wastes in question are taken back by the exporter or the generator or, if necessary, by itself into the State of export." On the other hand, if the illegal traffic is as the result of conduct on the part of the importer or disposer, "the State of import shall ensure that the wastes in question are disposed of in an environmentally sound manner by the importer or disposer."

As mentioned in the following sections, the states of export and import can often have different interpretations of how a shipment of waste is classified. Some shipments which have been regarded as hazardous waste by the state of import have not been regarded as hazardous waste by the state of export. Item 5 of Article 6 of the Basel Convention states that

> In the case of a transboundary movement of wastes where the wastes are legally defined as or considered to be hazardous wastes only: (a) by the State of export, the requirements on signing the movement document that apply to the importer or disposer and the State of import shall apply mutatis mutandis to the exporter and State of export, respectively; (b) by the State of import, or by the States of import and transit which are Parties, the requirements on notification, and allowance of export that apply to the exporter and state of export shall apply mutatis mutandis to the importer or disposer and State of import, respectively.

These definitions and measures are reflected in several guidance documents including the "Manual for Implementation" (Secretariat of the Basel Convention 1995a) and "Model National Legislation on the Control of Transboundary Movements of Hazardous Wastes and Other Waste and their Disposal" (Secretariat of the Basel Convention 1995b), which were approved by COP III in 1995. However, there are some unclear areas concerning illegal traffic in the Convention and guidance documents. For example, it is not clear how to identify the responsible actors of illegal traffic, although as noted above, the responsible actors (generator, exporter, importer, or disposer) should be identified in order to handle the illegally shipped waste in an environmentally sound manner. It is also not clear whose responsibility it is—the state of export or the state of import— for illegal traffic when there is a difference in interpretation of the hazard posed by the shipment.

9.1.2   Illegal Traffic Definition in National Legislation

Countries ratifying the Basel Convention have created national legislation to cover hazardous waste and other waste according to the definitions laid out in the Convention. In addition, each state can make its own regulations on non-hazardous waste and second-hand goods. Some examples are given in Chapter 1, and the national legislation enacted in China, Vietnam, and South Korea is explained in Chapters 3, 4, and 5, respectively.

Illegal traffic is divided into types, and this division depends on the structure of the national legislation. However, clear cases are as follows: if the import of specific recyclable waste or second-hand goods must satisfy a certain standard (e.g., regarding the quality of recyclable waste or age of the second-hand goods), illegal traffic would be a violation of such standard; if the import of specific recyclable waste requires an import permit issued by the government, import without permission constitutes illegal traffic.

The relationship between the Convention and national legislation is not so clear, however. Article 3 of the Convention states that each party shall notify other parties through the Secretariat of national definitions of hazardous waste that differ from those in Annex I and II of the Convention. Paragraph 2 of Article 13 of the Convention also require parties to inform each other through the Secretariat of (1) decisions made by the party not to consent totally or partially to the import of hazardous wastes or other wastes for disposal within the area under their national jurisdiction, and (2) decisions taken by the party to limit or ban the export of hazardous wastes or other wastes.

9.2   UNCOVERED CASES OF ILLEGAL TRANSBOUNDARY MOVEMENT OF WASTE

Although the Basel Convention and related regulations have been implemented, illegal transboundary movement of waste continues, and new cases are uncovered every year. This section describes a number of such cases.

9.2.1   Hazardous Waste and E-waste Shipment Seized in Hong Kong

Hong Kong is a duty-free port known as an international through-port for goods from Europe and the United States bound for China and the Asia region. It is reported that e-waste is exported to Guangdong Province in China through Hong Kong (Basel Action Network and Silicon Valley Toxic Coalition 2002).[2] Because of its geographical location and economic function as the gateway to mainland China, many shipments of hazardous waste, including electronic waste, pass through Hong Kong.

A common question asked is how is e-waste being imported into the

Table 9.1   *Transboundary movement of electronic waste between January 2006 and October 2008*

Exporting country	No. of convictions	Types of hazardous waste	Total fines and other penalties(HK$)
Japan	31	Batteries and cathode ray tubes	860,400
United States	26	Batteries and cathode ray tubes	710,000
Canada	14	Batteries and cathode ray tubes	353,000
Korea	10	Batteries and cathode ray tubes	330,000 and a community service order for 180 hours
Ghana	7	Batteries	75,000
United Arab Emirates	6	Batteries and cathode ray tubes	160,000
Other places[a]	44	Batteries and cathode ray tubes	949,000

Note: a. "Other places" include 24 other countries, including Guatemala, Malaysia, Singapore, and Italy, each with fewer than six related convictions.

Source: Press Release at Hong Kong Government News (news.gov.hk) http://www.info. gov.hk/gia/general/200903/18/P200903180161.htm.

mainland? On the American TV program "60 minutes" in November 2008, CBS Broadcasting Inc. reported that toxic electronic waste was being smuggled into Hong Kong in containers from the United States by thousands of merchant vessels every year, for storage and subsequent shipment to the largest hub of electronic waste in mainland China – the town of Guiyu in Shantou City. According to the Hong Kong Environmental Protection Department (EPD), between 2006 and 2008, over 900 inspections were conducted of suspicious storage sites and 13 joint raids with other law enforcement agencies were launched at sites involving illegal activities. Furthermore, the EPD completed the prosecution of 197 cases between January 2006 and October 2008. Among these 197 prosecutions, 138 convictions were made. Details of the types of e-waste and the countries of export in each case are shown in Table 9.1.

Between 2006 and 2008, 291 imported shipments of controlled electronic waste were returned to their countries of origin. Table 9.2 shows the number of shipments and the countries of export.

Figure 9.1 shows the number of shipments of hazardous waste import detected in Hong Kong from 2006 to 2010. In 2007, the number of

Table 9.2   Number of shipments of electronic waste returned for illegal import between January 2006 and October 2008 (no. of containers)

Countries of export	No. of shipments (No. of containers)[a]
United States	110 (140)
Japan	34 (39)
Canada	20 (30)
Vietnam	13 (45)
Australia	11 (15)
United Arab Emirates	11 (13)
Other places[b]	92 (139)

Notes:
a.   Some shipments involved more than one container.
b.   "Other places" include 40 other countries, including Guatemala, Algeria, Malaysia, the Philippines, and Singapore, each with fewer than eight related shipments.

Source: Press Release at Hong Kong Government News (news.gov.hk) http://www.info. gov.hk/gia/general/200903/18/P200903180161.htm.

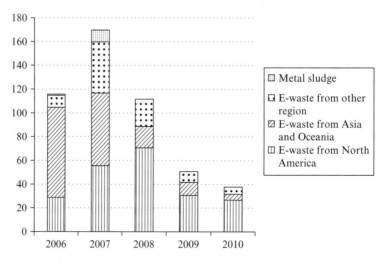

Source: Compiled from data from the Environmental Protection Department of Hong Kong.

Figure 9.1   Number of shipments of hazardous waste detected in Hong Kong from 2006 to 2010

detected shipments exceded 160, but the number has decreased significantly since. Especially, the number of detected e-waste shipments has decreased. Stricter enforcement of the import regulation discourages the traders from exporting e-waste to Hong Kong, with a possible final destination of mainland China. In this period Japan was a major exporting country of e-waste to Hong Kong. Based on the information provided by the Environmental Protection Department of Hong Kong, the Japanese Ministry of Environment issued an announcement on "Export of used CRT TV and monitors to Hong Kong" in June 2007. It stated that even if the goods are regarded as second-hand goods by the exporter, they may be regarded as hazardous waste by the Hong Kong government and that some shipment had been shipped back to Japan. This announcement forced traders to stop the export of e-waste from Japan to Hong Kong.

9.2.2 Medical Waste Exported from Japan to the Philippines[3]

In 1999, a container labeled as having waste paper and waste plastic contents, shipped from Japan to the Philippines, was abandoned in the port of Manila. The Philippine government investigated the container and found medical waste, and in accordance with the procedure defined in the Basel Convention, they requested the Japanese government to retrieve the waste shipment. The Japanese government complied and repatriated the waste to Japan in the form of an administrative subrogation and incinerated it. The president of the exporting company, Nisso Industry Co. Ltd was arrested, and both the president and the company were found liable for illegal waste shipments and have since been convicted. However, the suspects were not prosecuted in line with the Basel Convention or the waste disposal law, but in accordance with the Foreign Exchange Act (under the unmoderated exports provision). The waste was not regarded as hazardous because no hazardous substances or infectious materials were definitively found despite a rigorous investigation.

9.2.3 Used Tire Imports from the United Kingdom to Thailand[4]

In 2002, 23.5 tons of used tires were imported from the United Kingdom to Thailand. However, after arriving in Thailand, the importer did not come to collect the containers within the stipulated 75 days. Therefore, the containers were opened by customs and used car engines and used electric appliances were found in addition to the used tires. An exporter in the United Kingdom was identified as being in violation of the Basel Convention. The exporter subsequently reclaimed the batteries and electric appliances, but the waste tires were not returned because they were not defined as hazardous waste. In May 2003, the Government of Thailand promulgated restrictions on the import of used tires.[5]

Ratifying parties were informed of this regulation by the Basel Convention Secretariat. Although used tires in Thailand are still not considered hazardous waste, Thailand is applying more stringent regulations on the import of such goods in an effort to protect health and the environment.

9.2.4   Plastic Waste Exported to Thailand from the Netherlands

In 2003, the Netherlands exported 17.7 tons of material described as high-density polyethylene and low-density polyethylene on the application form. After arriving in Thailand, the importers did not collect the shipment within the 75-day period and it was therefore opened by customs, who discovered that the material was actually imported waste paper and waste plastic.

In Thailand, the import of waste plastics for recycling requires a permit from the Department of Industrial Works (DIW) and must comply with the procedures specified in the notice of the Ministry of Industry on Standards for the Import of Materials consisting of waste plastic in 1996[6] and Notification No. 112 of the Ministry of Commerce for Imports in 1996.[7] However, the procedure for obtaining an import permit had not been followed for the imported material. The Pollution Control Department (PCD) reported to the Competent Authority of the Netherlands and the waste was returned. Dutch authorities investigated the case and found that the company that exported the waste had already closed down. This case is an example of import under false declaration. Because import restrictions on plastics have since been transmitted to the Basel Convention Secretariat and all ratifying parties, waste can now be shipped back to the exporting country.

9.2.5   Export of Used Pachinko Machines from Japan to Thailand

Used pachinko machines, which are a popular game machine in Japan, were imported to Thailand in 2004 (Figure 9.2). In September 2003, Thailand had applied import restrictions on used electronic appliances, including those imported with the correct permit. However, the importer of the used pachinko machines did not obtain the necessary import permit; therefore, return measures were taken. In this case, the shipper may not have been aware of the new regulation because the shipment was made just after the regulation was implemented.

Source: Courtesy of Ms Teeraporn, PCD.

Figure 9.2   Used pachinko machines exported from Japan

Restrictions on the import of used electronic appliances by Thailand were issued in September 2003 by the DIW.[8] The regulation was most likely implemented because of the possibility of increased imports of used electronics as a result of a Chinese regulation in 2002 prohibiting the import of used electronics.[9] This case is an example of regulation transfer from one country to another (Jänicke 2006).

The Thai regulation covers 29 items including appliances such as PCs and TVs and their constituent parts, and is similar to the Chinese regulations. However, if imported items can be sold and reused, electrical and electronic equipment used within three years from its date of manufacture (copy machines within five years) are allowed to be imported. If the importer wants to dismantle and recycle the item, import of used electronics is allowed under the following conditions: (a) the importer must be a factory registered with the DIW; (b) the imported items must have economic value; and (c) the exporting country must be party to the Basel Convention. These points differ between the Chinese and Thai regulations. China prohibits importing waste electrical and electronic equipment. The second-hand goods must have the same quality testing as new products.

It is very difficult to conduct such testing for every type of second-hand goods. As a result, the testing requirement can be regarded as non-tariff barrier on the import of used electrical and electronic equipment.

The Thai regulations require importers to submit their application with the packing list and a letter guaranteeing that the imported goods were manufactured within three years at each time of importation. In addition, the DIW conducts inspection of importers more frequently than in the past. According to an importer of used Japanese home appliances, the regulations force unofficial (unregistered) importers out of business. The DIW issued import permits to 53 companies in 2006. The volume of shipments handled with import permits reached 216,932 tons in 2004, 217,467 tons in 2005, and 216,817 tons in 2006.

9.2.6   Plastic Waste Exported from Japan to Qingdao, China

In late March 2004, Chinese authorities discovered around 4,000 tons of waste plastic bound from Japan to Qingdao, China, in violation of regulatory standards. According to reports in China, the shipment from Japan tried to conceal large amounts of contaminated plastic with a small covering of good quality waste plastic. The imported contaminated plastic would have been illegally resold to farmers who do not have permission to process imported plastic.

On May 8, 2004, the import of waste plastic to China from Japan was banned by the Chinese government who took a very serious view of the incident (Directorate of Public Notice No. 47 on National Quality Supervision Inspection and Quarantine). Banning the import of a recyclable waste from a specific country was the first disciplinary action taken by the Chinese government. The Chinese central government requested the Japanese exporter to (1) return the entire shipment of waste plastic to Japan, (2) compensate the buyers of the waste plastic, and (3) requested the Japanese government to take measures to prevent similar shipments. These requests were the condition for resuming the import of waste plastic from Japan. The Chinese government regarded the imported contaminated waste plastic as hazardous waste and therefore a controlled material under the Basel convention. However, the Japanese government did not consider it as hazardous waste because the shipment did not contain hazardous substances. The two countries were deadlocked in their conflicting views.

Because of a prolonged official investigation and trial, as well as time-consuming measures such as compensation for the buyers, reclamation of the waste plastic was not started until September 2005. Part of the shipment has been confirmed to have returned to Japan, but most

Table 9.3   Waste returned to Korea from China

Year	Number of exports (A)	Amount (million tons)	No. of return shipments (B)	Ratio of returns (B/A)	Related e-waste
2005	4,750	58.2	28	0.59%	4
2006	5,081	54.9	12	0.24%	N/A

Source: An Hong Jun (2006) and the Korea Zero Waste Movement Network (2006).

of it was distributed to unknown destinations. The chairman of the importer was sentenced to ten years in prison and fined 200,000 CNY, while a fine of 800,000 CNY was imposed on the importer and processing company.

9.2.7   Return Shipment from China to Korea

China conducts inspections before accepting shipments from an exporting country in efforts to prevent illegal waste imports. In Korea, since 1999, Pan-Kan Test Co. has been carrying out pre-shipment inspection services involving radioactivity testing with portable detectors and visual inspection to discriminate between acceptable and export-prohibited items. When using a Pan-Kan test service, the exporter must be officially registered with the Chinese government as a waste exporter. In 2007, about 300 companies registered.

According to Pan-Kan Test Co., China deemed 28 cases in 2005 and 12 cases in 2006 to be illegal exports and shipped them back (Table 9.3). Of these 28 cases, 4 were returned under the category of e-waste. The returned waste included printers, fax machines, copying machines, odorous goods (wood/shoes/gloves), video tapes, compressors, computer monitors, plastics, and medical waste.

Following the Lehman Brothers shock of 2008, the export volume of recyclable waste decreased by around 20 percent. However, as of late 2009, it had recovered to previous levels. As for the number of returns, seven to eight cases are reported annually, accounting for approximately 0.1 percent of the total number of exports in 2008 and 2009. Specifically, in April 2009, mixed fiber cable was discovered in 90 tons of recyclable waste headed for Tianjin. In March 2009, 20 tons of recyclable vinyl going to Qingdao was returned because of its soiled appearance. Additionally, a case of illegal export was also detected, but not through the abovementioned Pan-Kan testing. Nine companies in the Inchon region of Korea were exposed for exporting waste without following the required procedure, chiefly printed circuits (*Bupyong* newspaper, March 2010). They had exported 900 tons of e-waste between November 2008 and the end of 2009.

There are two main ways to treat the handling of hazardous waste after return: in Korea as the exporter's responsibility, or in a third country for re-export. Although Pan-Kan Test Co. serves primarily to verify waste exports, it does not take any direct responsibility for return because approval by Pan-Kan Test Co. does not guarantee export permission by the Chinese government.

According to the Act on the Transboundary Movement of Waste and its Treatment (ATW), returned waste can be treated as legal execution by proxy. However, this applies only to exports without approval from the Ministry of Environment (MOE) and the possibility of unexpected pollution. In addition, it is also required that the exporter does not have the financial ability to treat the hazardous waste. In other words, Korea does not have any provisions that cover returned shipments of waste. Until now, it has not been reported whether the MOE is involved in the treatment of return waste from China.[10]

9.2.8   Compost Exported from Singapore to Indonesia[11]

In 2004, the local environmental management office of the Batam Islands in Indonesia received notice of a sighting of something resembling hazardous waste on Galang Baru Island. The material was 1,149 tons of compost imported from Singapore on July 28, 2004. Inspection by the Ministry of Environment of Indonesia found that the imported compost was in fact hazardous waste because it contained a high concentration of metals. In discussions between the Indonesian and Singaporean governments conducted in late 2004, the Singaporean side claimed the waste was not hazardous and was simply compost. In March 2005, in response to the conflict, someone daubed graffiti on the gates of the Singaporean embassy in Jakarta.

In May 2005, the Singaporean and Indonesian governments reached an agreement, which was witnessed by the Basel Convention Secretariat in Geneva. The imported material could not be regarded as hazardous waste according to the regulations of Singapore and the Basel Convention, but it was considered as hazardous waste according to the regulations of Indonesia. However, the restriction of hazardous waste in Indonesia was submitted to the Basel Convention Secretariat on July 29, 2004 and the information was disseminated to ratifying parties including Singapore on August 27, 2004. When the compost was shipped, the Singaporean government had no responsibility to control the material. Therefore, in order to resolve the dispute, the Singaporean government agreed to issue a permit to retrieve the goods.

9.2.9 Export of Second-hand Lead-acid Battery Waste from Japan to Vietnam and Hong Kong

A sharp increase in nominal exports of second-hand lead-acid battery waste from Japan has been observed since early 2005 (Figure 9.3). According to trade statistics, the HS code (Harmonized Commodity Description and Coding System) for second-hand lead-acid batteries does not distinguish them from new lead-acid batteries. Therefore, the exact quantity of export of second-hand lead-acid batteries is unknown. However, based on data from the Ministry of Economy, Trade and Industry's prior consultation service and a review of the trade statistics, most of the batteries exported to Vietnam and Hong Kong are believed to be waste lead-acid batteries.

It is not clear whether the exported lead-acid batteries were reused or not. The Japanese Ministry of Environment and Ministry of Economy, Trade and Industry issued a notice in April 2006 for preliminary consultation relating to the export of lead-acid batteries. This measure requested that during the prior consultation service, exporters provide pre-shipment verification of direct re-use and the existence of users at the destination. The pre-shipment inspection includes verification of "a selection of functioning items," "no damage to the battery casings," and "testing of power-on prior to exporting, as well as removal of non-working items."

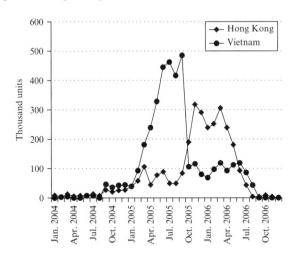

Source: Compiled from Japanese Trade Statistics.

Figure 9.3   Export of lead-acid batteries from Japan to Hong Kong and Vietnam

Source: Photo by Michikazu Kojima (January 2007).

Figure 9.4   Waste lead-acid batteries exported from Japan in Guangzhou, Guangdong, China

In addition, the name and address of the buyer in the exporting country should be provided during the consultation service.

Because it was not confirmed that the exported items were re-used, Vietnam and Hong Kong regarded the items as hazardous waste and have enhanced their inspections. Since the imported batteries seemed to have been re-exported to other countries, the Vietnam government issued a notice to enhance import controls for regulating importation of waste for re-export purposes. In 2006, the Japanese Ministry of the Environment and Ministry of Economy, Trade and Industry issued a joint notice to some exporters as the Hong Kong and Vietnam authorities had ordered a return shipment of some containers with damaged lead-acid batteries.

The export of waste batteries has decreased rapidly under stricter trade regulations and customs inspections. However, some sources have pointed out the possibility that some exporters might have shipped waste lead-acid batteries concealed with mixed metals. Although it occurred after trade statistics showed a decrease in the export of lead-acid batteries, a pile of waste lead-acid batteries with Japanese instructions was found in a small-scale recycling factory in Guangzhou, China in January 2007 (Figure 9.4). The recycling factory in question dismantled the batteries, shipped the lead electronic plates to another recycling company, and melted the terminals into ingots.

The export of waste lead-acid batteries to South Korea as hazardous waste, which goes through prior notice and consent procedures, has increased after the enforcement of the regulation become stricter. In 2005, the amount of export of waste lead-acid battery to South Korea was 4,127 tons. It reached 46,614 tons in 2007 and 79,996 tons in 2009.

9.2.10 Second-hand Automobile Parts Exported from Japan to Africa

In 2006, second-hand auto parts were exported from Japan to Africa, via France, and the container was intercepted and examined by French authorities. The French government regarded the second-hand auto parts as hazardous waste in accordance with the Basel Convention because the packaging was not appropriate to prevent damage and oil leakage during transportation. During consultation between the French and Japanese governments, the Japanese government claimed that the container did not qualify as hazardous waste under the Basel Convention. Although discussions were deadlocked, the exporter was forced to notify the French government, which permitted the export in accordance with European Union (EU) regulations.

9.2.11 Waste Plastics Imported from the United Kingdom to China

In January 2007, 200,000 tons of plastic waste and 500,000 tons of waste paper were imported from England to Foshan district in Nanhai, Guangdong Province. According to a survey conducted by the Guangdong Environmental Protection Bureau (EPB), Lianjiao village and an industrial park in Nanhai have become a well-known regional center for the trading and processing of domestically generated waste plastic since the late 1970s. Annually, the region recycles over 20,000 tons of waste plastic, of which more than 80 percent was collected domestically. However, 90 percent of imported waste plastics have been imported without the required waste import permit from the State Environmental Protection Administration (SEPA).

According to a report by China Central Television, the Nanhai City EPB expressed the intention to eradicate the waste plastics recycling industry in the region and also requested the return of approximately 700,000 tons of imported waste from the United Kingdom. The SEPA and Guangdong Province EPB have stated that the regulation will be applied strictly to imports of recyclable waste.

According to an interview with a British official at the Sixth Session of Open-ended Working Group of the Basel Convention,[12] the amount of recyclable waste exported from the United Kingdom to China has increased 158-fold over the past eight years. British authorities have also participated in the European Union Network for the Implementation and Enforcement of Environmental Law (IMPEL) to try to prevent illegal shipments. However, it is difficult to guarantee that exported recyclable waste meets the Chinese environmental standards.

9.3   POINTS OF COMPARISON AMONG THE CASES

The previous section explained various illegal shipments involving Asian countries. This section summarizes the characteristics of each case from several standpoints. Table 9.4 summarizes the cases mentioned in the previous section.

9.3.1   Countries Involved

This chapter deals only with cases involving importing or exporting Asian countries. All Asian countries involved in the case studies are party to the Basel Convention. On the other hand, some hazardous waste imported into Asia originates from the United States, which is a signatory of the Basel Convention, but not a ratifying party, and has its own regulations on the transboundary movement of hazardous waste. Thus, illegal shipments can occur even under the Basel Convention.

In addition, some illegal shipments are originating in less developed countries and sent to middle-income developing countries. For example, Hong Kong uncovered waste batteries and CRT from Ghana (see Section 9.2.1), where disposal and recycling facilities with environmentally sound management may not exist. As mentioned in Section 9.1, illegal shipments as a result of the exporter's conduct should be taken back by the exporter, generator, or state of export. The Basel Convention is formulated without taking into account that developing countries can sometimes send hazardous waste to other developing countries.

9.3.2   Types of Waste and Interpretation of Definitions

The types of waste concerned include e-waste, lead-acid batteries, hazardous industrial waste, mixed waste, and tires, among other materials. Lead-acid batteries contain hazardous substances, whereas waste tires or waste plastics are not usually hazardous. Because the demand for waste tires is not large, their value is usually very low or even negative. Therefore, there

Table 9.4 Summary of the case studies presented

Year	Importing and exporting countries (status of ratification)	Declaration in the exporting country	View of the importing country on imported material	Content
Sept 1996– 2005	Asia, North America, Europe, and Africa (party and non-party) → Hong Kong (a region of China, which is party)	Mixed metal, plastic waste, second-hand electronics, etc.	Medical waste, cathode ray tubes, hazardous waste, and other waste	From 2006 to 2008, the number of waste return shipments reached 291
1999	Japan (party) → Philippines (party)	Used paper and waste plastics	Mixed hazardous and medical waste	After returning to Japan, contamination by toxins was tested for, but no toxic substances were clearly found
2002	United Kingdom (party) → Thailand (party)	Waste tires	Stacked electric appliances, batteries, and other waste were returned. There is no demand for waste tires	Electric appliances, batteries, and other waste were returned. There is no demand for waste tires
2003	Netherlands (party) → Thailand (party)	High-density polyethylene and low-density polyethylene	Plastics waste	Violation of Thai regulations Transaction requires a permit
2004	Japan (party) → Thailand (party)	Used pachinko machines	Used pachinko machines	Violation of Thai regulations. Transaction requires permit. Returned
2004	Japan (party) → China (party)	Waste plastic	Waste plastics with garbage from daily life	Japanese government regarded it as non-Basel waste, but export company clearly violated Chinese regulations

Table 9.4 (continued)

Year	Importing and exporting countries (status of ratification)	Declaration in the exporting country	View of the importing country on imported material	Content
2005–2009	Korea (party) → China (party)		E-waste, odorous items such as shoes, gloves, and medical waste	In 2005, 28 cases of return shipments for Korea and 12 additional cases were reported
2004	Singapore (party) → Indonesia (party)	Compost	Hazardous waste containing heavy metals	Indonesia and Singapore differed in their views. Singapore permitted imports from Indonesia and did not regard the shipment as hazardous waste
2005–2006	Japan (party) → Hong Kong (party), Vietnam (party) → China?	Second-hand lead-acid batteries	Waste lead-acid batteries (hazardous waste)	Return shipment measures were taken by Hong Kong and Vietnam
2006	Japan (party) → France (party) → Africa (not clear)	Second-hand car engines	France regarded it as hazardous waste.	Japan regarded waste items as second-hand goods. France issued a report permit to Africa
2007	United Kingdom (party) → China (party)	Plastics, paper, etc.	Waste from daily life	The case is regarded as a violation of domestic law. Returned to exporting country
2009	Africa → Malaysia (party) → Singapore (party)	N/A	N/A	Waste generator and trader not identified

Source: Compiled from various sources.

is a possibility that low-demand non-hazardous waste is being intentionally dumped in developing countries.

As noted earlier, in some cases the views of the exporting countries on specific items differ from those of the importing countries or transit countries. For example, compost shipped from Singapore was regarded as hazardous waste in Indonesia; second-hand automobile parts shipped from Japan were regarded as hazardous waste by the French government. In neither case were the items viewed as hazardous by the exporting countries. These kinds of issues occur due to ambiguity in the definition of hazardous waste and other terms including product, second-hand goods, and non-recyclable waste.

Honda (2009) defines illegal transboundary movement as the transboundary movement of waste controlled by both the exporting and importing parties beyond the jurisdiction of the Basel Convention, while illicit transboundary movement is the transboundary movement of gray-area waste controlled only by the importing party, but again beyond the jurisdiction of the Basel Convention. This argument is not so clear however. In the Basel Convention, illegal traffic is defined in Article 9, discussed in Section 9.1. Even if a material is defined as hazardous waste by the importing country but not the exporting country, transboundary movement of hazardous waste without notification and consent is regarded as illegal traffic. On the other hand, in the case that the material is non-hazardous waste, which is beyond the jurisdiction of the Basel Convention, responsibility for illegal traffic of the material may not to be needed to follow the procedure of the Basel Convention.

However, there is the possibility that the exporting country will regard a shipment as non-hazardous or goods but the importing country will regard it otherwise; the "compost" exported from Singapore to Indonesia is an example. As seen above, the reporting of national definitions to other parties through the secretariat can play a crucial role in resolving conflicts between the exporting and importing country.

9.3.3   Prior Notice and Consent

None of the cases presented in Section 9.1 went through the procedure of prior notice and consent. It is not clear, therefore, whether exporters and importers were aware of such regulations. However, both exporters and importers of waste lead-acid batteries are likely to conduct international trade intentionally under the guise of second-hand goods. In such cases, exporters or importers use fake documents for shipment. On the other hand, exporters of automobile parts from Japan considered the shipment to be legal because such parts are typically reused.

9.3.4   Who Should be Responsible?

It may be more useful to classify actual cases in terms of responsibility of the waste generator, exporter, importer, and disposer. As mentioned in Section 9.1, the responsibility for handling illegal shipments is placed on the waste generator/exporter or importer/disposer, based on the fact that the illegal traffic is the result of conduct on the part of the waste generator/exporter or importer/disposer. However, it is not clear what kinds of conduct are considered to violate the regulation. If the importer tells the exporter that the traded material is not regarded as hazardous waste in the importing country, but in fact the traded material is regarded as hazardous waste by the government of the importing country, then who should be responsible for the ship-back?

Various scenarios should be analyzed in terms of responsibility of the waste generator/exporter and the importer/disposer, as follows.

(1)   Both parties regard the traded material as hazardous waste.
(2)   The importing country regards the traded material as hazardous waste, while the exporting country does not. The importing country has submitted their definition to the Secretariat of the Basel Convention.
(3)   The importing country regards the traded material as hazardous waste, while the exporting country does not. The importing country has not submitted their definition to the Secretariat of the Basel Convention.
(4)   The importing country regards the traded material as hazardous waste, while the exporting country does not. The importing country has not submitted their definition to the Secretariat of the Basel Convention. Additionally, the importer/disposer told the waste generator/exporter that the traded material is not regarded as hazardous waste in the importing country.

9.4   MEASURES TO PREVENT AND HANDLE ILLEGAL SHIPMENTS

Several possible trade measures have been designed to prevent illegal trade of hazardous waste. To consider appropriate policies, the difference between illegal and illicit trade should be clearly taken into account.

9.4.1   Basel Ban Amendment

Many parties have expressed hope for the early entry into force of the Ban amendment to reduce improper transboundary movements of hazardous waste. However, the Ban amendment will not contribute to reducing the number and type of cases like those mentioned in Section 9.2, because prior notice and consent was not used in any of these cases. If the exporter is trying to export hazardous waste under the guise of products, second-hand goods, or non-hazardous recyclable waste, the government may not be aware of the shipment of hazardous waste. More effort to enforce the law, such as frequent inspection of cargo, is the only measure that will reduce such illegal trade. On the other hand, the Ban amendment will affect official transboundary movement of hazardous waste, but may not prevent illegal movement; thus it is not effective for reducing the illicit trade of hazardous waste.

9.4.2   Information Sharing among Parties

To reduce illegal traffic due to misunderstanding of regulations, exporters and governments of exporting countries should be aware of existing trade regulations and standards on hazardous waste, second-hand goods, products, and non-hazardous recyclable waste in the importing country.

The Basel Convention has schemes to facilitate information sharing, which are defined in Articles 3, 4, and 13. Parties can disseminate the national definition and list of hazardous waste and other waste, as well as prohibit the import of hazardous waste, through the Basel Convention Secretariat. Although the information-sharing scheme is defined in the Basel Convention, only a few countries utilize the mechanism. Even though the definition and list of hazardous wastes have been disseminated to other countries, ambiguity in the definition might still exist. Furthermore, the exporter or the government of the exporting country may receive a different interpretation from the government of the importing country.

In addition, some shipments exported from the United States, which is not party to the Basel Convention, were seized in Hong Kong. In response, Hong Kong has developed necessary procedures in accordance with the Basel Convention. However, the regulations of Hong Kong are not identical to those of China. Hong Kong cannot disseminate its regulations through the Secretariat of the Basel Convention to other parties because China's Ministry of Environment is the competent authority.

It is very important for all parties to submit their definition and list of hazardous waste to the Secretariat of the Basel Convention. It may be useful to more clearly specify certain hazardous waste by identifying sub-categories, such as certain types of waste printed circuit board (PCB), for example PCB with IC-tips and PCB without IC tips. Regarding the sub-categories, a questionnaire should be sent to those party to the Basel Convention as well as those not party to it and the responses shared among parties.

9.4.3 Information Dissemination in Exporting Countries

Although importing countries have sent a clear definition and list of what they regard as hazardous waste to other parties through the Basel Convention Secretariat, exporters may not be aware of them. Information dissemination efforts in the exporting countries are thus another key factor in reducing illicit transboundary movement of hazardous waste.

A good example of information dissemination practiced by the Japanese government is the organization of seminars more than ten times every year (Yoshida and Kojima, 2008). In the seminars, the Japanese Basel Law has been explained and lectures given. Moreover, the Japanese government has provided consultation services to exporters.

9.4.4 Inspection of Shipments

Inspection at ports is key to preventing illegal shipments. Exporters sometimes forge shipping documents and send hazardous waste overseas. Because of budget constraints, however, it is very difficult to check every shipment. However, based on the various information collected by and shared among different government agencies, inspection should be conducted effectively. To institutionalize such information sharing, it is good to establish a committee for enforcement of the Basel Law. The committee should consist of ministry staff in charge of the environment, industry, customs and others.

9.4.5 Ship-back

From the case studies, it is obvious that the ship-back procedure is not standardized. Based on the procedure defined in the Basel Convention, the starting point in shipping back hazardous waste is to identify who is responsible for the illegal traffic, the waste generator/exporter or the importer/disposer. As mentioned in Section 9.3, however, there is no guidance document to identify the responsible actors. For example, if the government of the importer found the waste shipment to violate regulations in its country but not in the exporting country, and if the importer regarded it as non-hazardous waste and the exporter did not know of the importing country's regulation, who should be responsible? If the exporter insists that the waste in the container is not the original contents, and if there is the possibility that some other party secretly substituted recyclable waste into hazardous or non-recyclable waste, who should be responsible? There is obviously a need for a formal procedure to identify the responsible party more clearly.

The process of how to implement ship-backs is also unclear. The EPD in Hong Kong considers putting responsibility on the transportation company is a practical way to implement ship-backs. While this is true, it is not the same as the procedure set out in the Basel Convention. There is also the possibility that ship-backed waste is redirected to another country, instead of being returned to its point of origin. A traceability system for ship-backed waste should be developed. A mechanism for the disclosure and sharing of information on ship-backed waste between the relevant governments may provide a starting point for developing proper management of ship-backed waste.

9.4.6   Proper Treatment without Ship-back

If the country of origin of illegal traffic has no facility to treat the hazardous waste in an environmentally sound manner, ship-back may not be the appropriate measure. In Asian countries, some shipments were found to have originated in Africa, where there are few facilities with environmentally sound technology.[13] Paragraph 2 of Article 9 of the Basel Convention states that if the exporter, generator, or state of export cannot take back the waste, the shipped waste should be disposed of in accordance with the provisions of the Convention. Probably this means that the importing country should treat the imported hazardous waste, if possible. However, preventive measures should be implemented in the state of export to prevent similar shipments in the future. Some kind of mechanism to improve the state of export's management system should be considered.

9.5   CONCLUSION

This chapter has presented a number of illegal and illicit shipments of hazardous waste uncovered in recent years. To determine appropriate policy, it is important to understand the background of these cases. Examination of various cases helps to clarify the form that appropriate countermeasures and policy for preventing illegal and illicit transboundary shipments of hazardous waste should take. Hazardous waste should also be defined and listed clearly by each party, and the information should be disseminated to other parties though the Basel Convention Secretariat. There is also the need to establish mechanisms for monitoring and enforcement as well as managing waste afterwards. The same information should also be disseminated to exporters by the governments of exporting countries, and enforcement should be strengthened to prevent illegal export of hazardous waste.

NOTES

1. "Death toll from Ivory Coast pollution rises to 15," Agence France-Press, dated February 17, 2007. See also "The Probo Koala and the Ivory Coast" dated October 26, 2006, Trafigura website.
2. See also "Village resists arrival of foreign garbage" on the website of China's State Environmental Protection Administration, accessed on January 18 2007, http://english.sepa.gov.cn/zwxx/hjyw/200701/t20070118_99774.htm.
3. Japan Environmental Council (2002) and Tsuruta (2005).
4. In Thailand, the Department of Industrial Works is the Competent Authority for the Basel Convention, while the Pollution Control Department is the focal point for the Convention. The information on cases of import to Thailand in this chapter was collected in an interview with Ms. Teeraporn Wiriwutikorn of the Pollution Control Department and newspaper articles.
5. Goods classified by HS Code 4012.11, 4014.12, 4012.199, 4012.209, and 4004.00 are restricted.
6. Notification of the Ministry of Industry on the criteria for the approval of the scrap and used material which is made of used and unused plastics, 1996.
7. Notification of the Ministry of Commerce on the import of goods No.112, 1996. It stated that the objectives of import restriction were to protect the environment and to prevent hazards to consumers' health. The import tariff on waste plastics was increased to 30 percent under the measure. It is significantly higher than for other scrap, for which the import tariff is only 1 percent.
8. Notification of the Department of Industrial Works on the Criteria for the Approval of the Import of Used Electrical and Electronic Equipment into Kingdom of Thailand, 2003.
9. See Chapter 3 of this book and Chapter 2 of Kojima ed. (2005).
10. This information is based on interviews with a representative of the MOE in October 2009. The MOE takes the stance that waste imports and exports are basically commercial contracts between private parties; therefore, in the event of return shipment, hazardous waste should be treated as the responsibility of the exporter.
11. Based on the following: "Singapore insists matter exported to Indonesia not poisonous," LKBN Antara, dated November 1, 2004; the Press release of the Secretariat of the Basel Convention "Statement jointly adopted by the Republic of Indonesia and the Republic of Singapore," dated May 12, 2005; and interview with the Environmental Impact Management Department in Batam, Indonesia in December 2004 by M. Kojima.
12. The interview was conducted by M. Kojima.
13. A shipment of imported hazardous waste was found in Singapore and was thought to have originated in Malaysia. However, an investigation by the Malaysian government found the container originated from Africa and was in transit in Malaysia. The faked document was attached to the shipment from Malaysia to Singapore by a middleman in Malaysia, who had since disappeared. The waste was treated in Singapore. (Based on interviews conducted with officers in Singapore and Malaysia in December 2011.)

REFERENCES

An Hong Jun (2006), 'Problems in waste management and response', Materials submitted to the National Audit (Korean).

Basel Action Network (BAN) and Silicon Valley Toxic Coalition (2002), 'Exporting harm: the high-tech trashing in Asia'.

Honda, Shunichi (2009), 'Overview of transboundary movement of hazardous waste and other wastes in Asia', presented at The Asian Network Workshop.

Jänicke, Martin (2006), 'The environmental state and environmental flows: the need to reinvent the nation state', in Gert Spaargaren, Arthur P.J. Mol and Frederick H. Buttel, *Governing Environmental Flows – Global Challenges to Social Theory*, Massachusetts Institute of Technology.

Japan Environmental Council (2002), *State of Environment in Asia 2002/03*, Springer.

Kojima Michikazu (ed.) (2005), *International Trade of Recyclable Resources in Asia*, Institute of Developing Economies.

Korea Zero Waste Movement Network (2006), 'Inclination of recycling marker in Korea and exports', material publicized to the public hearing for measures for illegal export of E-waste and response for activating domestic recycling held in Korea congress.

Secretariat of the Basel Convention (1995a), 'Manual for implementation', http://www.basel.int/meetings/sbc/workdoc/manual.doc.

Secretariat of the Basel Convention (1995b), *Model National Legislation on the Management of Hazardous Wastes and Other Wastes as well as on the Control of Transboundary Movements of Hazardous Wastes and Other Waste and their Disposal*, http://www.basel.int/pub/modlegis.pdf.

Secretariat of the Basel Convention (2002), *Guidance Document on Transboundary Movements of Hazardous Wastes Destined for Recovery Operations*, United Nations Environment Programme.

Tsuruta, Jun (2005), 'Kokusai kankyou wakugumi jouyaku ni okeru jouyaku jissen no doutai bunseki – 1999nen sangyou haikibutsu yushutujiken wo sozain-ishite' (Dynamics of Practice Process Framework Convention on international environmental treaties: 1999 exports to industrial waste materials annually to the incident), in Yamamoto Takashi Shiroyama Hideaki (ed.), *Kankyou to seimei* (Life and Environment), University of Tokyo Press (Japanese).

Yoshida, Aya (2006), 'Chugoku ni okeru kaden no riyuusu risaikuru' (Electronics reuse and recycling in China), *C & G*, 54–9 (Japanese).

10. From shipbreaking to ship recycling: relocation of recycling sites and the expansion of international involvement

Tadayoshi Terao

INTRODUCTION

A large ship contains significant quantities of useful metals, especially iron (in the form of steel). Several decades after a ship is launched, its useful life ends and it is scrapped, supplying many tons of recyclable metal. With so much metal and such a long life, the recycling of a ship is more akin to the recycling of an entire building than the recycling of smaller objects with shorter lifespans such as home electronics.

From the Bronze Age onwards, metal has been recycled as much as possible. Iron is comparatively easy to obtain from ore, but the process requires a blast furnace. Iron is much easier to recover from scrap iron, which has thus become a very important raw material for iron manufacturing processes. Only a few low-income developing countries have blast furnaces today, so these countries depend on recycled scrap iron for all their domestic iron production. Most developing countries do not have sufficient scrap iron from sources such as old buildings and infrastructure, as they lack the prior economic development in which such items would be constructed with iron.

In most developing countries that are undergoing rapid economic growth, demand is increasing for iron and steel products, for use in both public infrastructure and private construction. Foreign currency restrictions make it difficult for developing countries to import all the iron and steel scrap they need, and in many cases the domestic steel manufacturing industry is also underdeveloped. The shipbreaking industry produces large quantities of iron. Iron from shipbreaking is more homogeneous in quality than ordinary scrap iron, and it is mostly in the form of plates, which are often ready for immediate re-use as construction material, or require minimal reprocessing costs.

The world's shipbreaking industry has shifted from East Asian countries such as Japan, Taiwan, and South Korea to South Asian countries such as India, Bangladesh, and Pakistan, as each country follows the curve of economic development. Recycled steel from shipbreaking continues to play an important role in the economic development of the countries where it is practiced.

The shipbreaking industry carries risks, however. After the shift of shipbreaking activity to South Asia from about 1990, problems such as industrial accidents and environmental pollution increased alarmingly, and calls arose for international management and regulation of the industry. International organizations have taken measures. For example, the United Nations Environment Program (UNEP) adopted the Basel Convention on the Control of Transboundary Movements of Hazardous Wastes and their Disposal (Basel Convention) and the International Maritime Organization (IMO) acted as a forum for shipping-related international debate. The International Labor Organization (ILO) has also expressed serious concern for labor safety.

This chapter starts with a brief history of the shipbreaking business since the mid-1960s. Section 10.2 discusses the main reasons why the industry has shifted between countries. Sections 10.3 and 10.4 describe the problems that have arisen since the early 1990s such as industrial accidents and environmental pollution. Sections 10.5 and 10.6 outline the Hong Kong Convention for the Safe and Environmentally Sound Recycling of Ships (the Ship Recycling Convention) adopted by IMO as an international measure to combat these problems, and discuss the importance of building a sound ship recycling system through institutional arrangements.

10.1 HISTORY OF LARGE-SCALE SHIPBREAKING AND IRON RECYCLING: FROM EAST ASIA TO SOUTH ASIA

This section briefly reviews the world history of the shipbreaking industry over the last 50 years. After the mid-1960s, shipbreaking boomed during times of high economic growth, when demand for cheap iron and steel articles increased, but information is unreliable until the 1970s—the earliest time for which statistics could be obtained on international trends in the shipbreaking industry, especially for the locations where the actual dismantling was conducted. Before that time, the only records available were very general, often no more than the flag states of the dismantled vessels. In this section, by examining the original data of Lloyd's Register, a source used by the most of the earlier literature, we review data on the country of demolition that could be extracted from 1967 onwards. The international movement of world shipbreaking from 1967 to 2008 is shown in Figure 10.1; only vessels of 100 gross tons (GT) or larger are counted as "ships" for this purpose.[1]

Analysis of the Lloyd's Register data for the period preceding the mid-1970s made it clear that Taiwan had overtaken Japan in the mid-1960s to become the world's largest shipbreaking country, remaining so until the end of the 1980s. Japan, Taiwan, and South Korea were joined in the 1980s by mainland China. Until the end of the 1980s, East Asia was the world's center of shipbreaking.

Through the mid-1970s, only 50,000 GT per year or less were being dismantled. Around that time, shipbreaking volumes increased, with an acceleration during the early 1980s to between 150,000 and 200,000 GT per year. From 1987 to 1990, however, there was a lull. Although a rapid increase began in 1992, the volume stayed below the 200,000 GT per year seen in the mid-1980s. Another lull occurred between 2004 and 2008, after which a new upward trend appeared.

Country-by-country analysis allows us to investigate the geographical movement of the shipbreaking industry. In 1967, Taiwan and Japan dismantled 739,000 GT and 796,000 GT, respectively, with Japan being the most active shipbreaking country in the world. In 1968, however, Taiwan's share increased rapidly to 1,133,000 GT to become the world leader, while Japan declined to 388,000 GT. Taiwan remained at the top until 1988.

We have not analyzed data prior to 1966, so there is currently no confirmation, but it is probable that Japan held the title of the world's busiest shipbreaking country. The United States was also active in the industry at a scale approaching that of Japan and Taiwan until 1972, as was Hong Kong. Taiwan, South Korea, and China had insufficient capacity to cope with the rapidly growing world demand for shipbreaking that started in 1982. Japan temporarily revived its activities, with shipbuilders trying to enter the shipbreaking market with the assistance of government subsidies, a measure taken to counteract the depression in the shipbuilding industry in the 1970s.[2]

World shipbreaking suddenly contracted in the latter half of the 1980s, because of decrease in the supply of ships for demolition, associated with increase in demand for maritime trade. When the industry started to recover in 1992, the major demolition countries had changed, with China the only remaining East Asia player. South Asian countries such as India, Bangladesh, and Pakistan took the lead in dismantling most of the world's large vessels. The trend fluctuated a little, with China taking the lead in 1993, just before the South Asian countries suddenly increased their

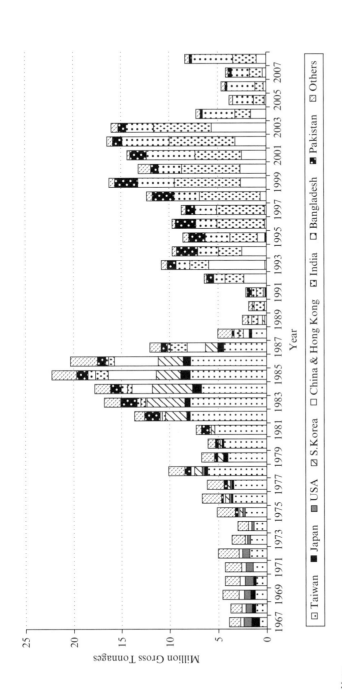

Note:
The data covers commercial ships with gross tonnage larger than 100.
Gross Tonnage (GT) is a unit based on the volume of vessel.

Source: Based on data from Shipbuilder's Association of Japan (eds) (2009), and Lloyd's Register, *Casualty Return*, various years.

Figure 10.1 Trend of world shipbreaking by major demolition countries (1967–2008)

shipbreaking activity. China kept even with the South Asian countries from 1999 to 2003.

As shown by Japan in the mid-1960s and Taiwan in the 1980s, the secure acquisition of a dismantling yard in a proper port was an essential condition for shipbreaking business development. In South Asia countries beginning in the 1990s, however, that condition changed drastically. In East Asia, countries like Japan, Taiwan, and South Korea, and many others established a wharf in ports where an initial rough disassembly was carried out using heavy equipment like large cranes. Dismantling parts of the ship into small iron plates and scraps was done later as part of a labor-intensive process. The most commonly used process in South Asia, called beaching, is entirely different from that in East Asia. In the "beaching" method, shipbreakers run the vessel aground at high speed and at high water, using a beach with no port facilities at all. Then the ship is cut into slices and pulled apart by human power alone, the only concession to modern technology being a winch whose wire can drag dismantled parts above the high-water line.

The South Asian practice is thought to have started when a large ship ran aground on rocks and provided rich pickings for the local scrap iron recovery business. The process had developed into the deliberate stranding of ships on beaches by the late 1980s. Shipbreaking is limited to a few areas, with most activity concentrated in a suburb of Chittagong, a major harbor city in the eastern part of Bangladesh, and in the Alan area of Gujarat State on the western coast of India. Although both areas have long beaches suitable for shipbreaking by beaching, many other areas have similar beaches, so natural conditions alone cannot fully explain the limited area of this practice. The next section discusses this issue further. Within South Asia, Bangladesh and Pakistan tend to disassemble larger ships than India does; a trait that has been constant ever since the early 1990s when shipbreaking started to be carried out extensively in South Asian countries.[3]

For any given maritime nation, there is a clear link between shipbreaking activity and economic development. In the mid-1960s, advanced nations such as the United States and rapidly developing ones such as Japan dismantled most ships, with Taiwan and Hong Kong close behind. The economy of Taiwan then started to expand and the country took over more than half the world's shipbreaking by the end of the 1970s. Rapidly increasing demand in the first half of the 1980s allowed South Korea and China to enter the shipbreaking industry.

Up until this time, shipbreaking became a brisk business wherever economic development was strong and demand for cheap steel was rapidly increasing to feed a growing construction sector. During the decline in world demand at the end of the 1980s, shipbreaking changed radically and in a way that would be difficult to reverse. With the advent of the beaching method in South Asia, which made dismantling much cheaper, former shipbreaking countries found it difficult to compete and have been unable to stage a strong revival of the business—all, that is, except China. Another reason for the shift of shipbreaking business is decline of domestic demand for low quality steel material in former shipbreaking countries as their economies grew rapidly. Although South Asian conditions are very different from those in East Asia, the cause and effect is much the same: demand for cheap steel during economic expansion leads to a rise in the demand for shipbreaking as a source of raw or part-processed materials.

10.2 SOURCES OF COMPETITIVENESS IN THE SHIPBREAKING INDUSTRY

The shipbreaking industry started because of its customers' need for scrap and iron plate. The sole source of raw material in the shipbreaking industry is the market for surplus ships. For large vessels, this market is international, and shipbreakers can obtain their raw material only by winning a competitive bid. In the early 1990s in the South Asian countries, shipbreakers had almost no fixed equipment, a cheap labor force, and no significant expenditures on worker safety or antipollution measures. These factors markedly reduced demolition costs. However, even these cost advantages did not justify bids higher than others in the international market, as profits depend on the demand for, and price of, the product, namely scrap metal, used machinery, and other reusable components. If the potential profit is high enough, it is worth bidding higher. Profits for shipbreakers are also increased if scrap buyers are not only willing to pay a high price, but are also located in the domestic market and close enough to keep transportation costs down.

Scrap steel plate, the main component of all ships, has special benefits for a developing country undergoing rapid economic growth. The plates are not only very large and usually of very high quality, but can also be processed without the use of an electric or blast furnace to melt them down. Plate can be reduced to steel bar simply by cutting it to the size required and then rolling it. This material is referred to as "ship plate." Even when reprocessing into new articles, ship plate is far cheaper to make than the same products whose steel comes directly from the blast furnace.

Although the ship plate business thrived in Japan and other developed countries until the 1960s with steel bar as the main output, it has almost disappeared now. The quality of ship plate depends entirely on the quality of the plate chosen when the scrapped ship was built, so stable output quality is hard to achieve. Most of the ship plate became bar steel used as the core of reinforced concrete, but in Japan the unpredictability of quality meant that articles made from ship plate could not meet the Japanese Industrial Standards (JIS) specification. Ship plate was no longer allowed in public construction and the ship plate business declined rapidly. The same phenomenon—a reduction in demand for cheap steel construction materials—is believed to be responsible for the decline of the shipbreaking industry in other developed countries.

Growing domestic demand for the iron and steel materials generated from dismantled vessels made the Taiwanese shipbreaking industry very prosperous by the mid-1980s. In terms of labor costs, Taiwan had no particular advantage over other developing countries, and like other countries, Taiwan also produced cheaply processed ship plate.[4] When shipbreaking started in South Asia, these countries, too, produced articles of iron or steel from ship plate for use in construction, and shipbreakers in South Asia have prospered since the early 1990s.

As we have seen, the South Asian beaching method lowers the cost of demolition sharply, using labor-intensive dismantling and low investment in equipment. Beaching certainly has problems with labor safety and environmental pollution, but it is a method that many developing countries, elsewhere than South Asia, can easily copy. The most important factor in determining the international competitiveness of the shipbreaking industry is domestic demand for cheap iron or steel materials. As many developing countries encourage economic development by protecting the domestic market so as to promote the domestic steel industry, the shipbreaking industry can take advantage of the price difference between the domestic and international markets for their products.

Taiwan's GDP growth exceeded 10 percent per year in the late 1960s, when it became the world's leading shipbreaking country. Domestic demand for steel construction materials was continuously high in Taiwan, but the import of steel was restricted significantly by tariffs and other non-tariff barriers. China Steel, a state-owned company, started integrated blast-furnace steel manufacture in the early 1970s, protected in the same way by the government.

The Shipbreaking Business Promotion Association, an incorporated foundation in Japan, found that high domestic demand for ship plate was one of the most important features of countries offering high prices for large scrappable vessels in the early 1980s, such as Taiwan, South Korea, and Pakistan; the domestic price for the plate made it a strong competitor with imported metal.[5] In Taiwan, high domestic prices for ship plate prompted the shipbreakers to bid high in the international market for their raw material. The protective Taiwanese tariff rate cannot fully account for this price differential between the domestic and international market. This difference suggests the existence of significant non-tariff barriers as well. At that time, the state-owned China Steel Corporation had started operating the first blast furnace in Taiwan, so the government's protective policy for the domestic iron and steel market may have played a significant role for shipbreakers too.

10.3 RESPONSIBILITY FOR RECYCLING: SHIP OWNERS OR SHIPBREAKERS

A large ship is a mobile object capable of crossing international boundaries. Its flag state does not necessarily indicate its geographical origin or the nationality of its owners, as many ships are registered under flags of convenience simply to reduce tax obligations. At the end of a ship's life, it is not possible to merely apply the rules to ships as for ordinary consumer durables such as home electronics, which can be moved across boundaries and then recycled. For home electronics, for example, the generator of the waste is responsible for correct processing of the good, a known object of a known nation. The same applies to automobiles. While an automobile can also move across borders, each automobile is registered in a specific country whose administration will manage its disposal. Assigning responsibility for the disposal of a ship on any individual or even any nation is more difficult.

Market supply for shipbreakers is unstable. Old ships provide the sole source of supply, and the definition of "old" varies according to the state of the market in international marine transportation. Most large vessels have a useful life of several decades, but it is not predefined. Life can be prolonged by repair or by change to less demanding uses (a luxury liner could be re-used as an inter-island ferry or pilgrim ship, for example). If there is an increase in maritime trade, it is not feasible to build new ships fast enough to satisfy demand, so ships that in normal times would have been scrapped are used longer and the supply of ships for dismantling dries up. If maritime trade suffers a setback, however, ship owners are reluctant to maintain ships that are not generating profit, so they send them for scrap and cause a market glut for the shipbreakers.

Instability is thus an integral part of the shipbreaking market. Significant changes have occurred in the market, not only in the numbers of vessels to be dismantled, but also in the countries that offer the service. For all the reasons outlined above, it is very difficult to design an international system for recycling. No country can be held responsible for the vessel itself, and each demolishing country sets its own rules. These difficulties suggest the necessity of building an international framework outside the Basel Convention.

10.4   PROBLEMS WITH LARGE-SCALE SHIPBREAKING AND PROVISIONS OF INTERNATIONAL TREATY

Even before the beaching method became common in South Asia, ship-breaking was labor-intensive and workers operated in a dangerous labor environment where serious accidents, including loss of life, occurred frequently. In the dismantling yards of Kaohsiung, Taiwan's main demolition port, a large explosion killed many workers in 1986 and was a major factor in the discontinuation of demolition in Kaohsiung. Even Kaohsiung, however, was much safer, and much less polluting, than beach demolition.

On sand, all the waste oil and other toxic fluids that arrived with the vessel will leak directly into the surrounding environment. The ship's abundant asbestos may harm laborers' health, and workers are not issued adequate protection against that or any other hazards, so they face grave risk of accident and illness. Even when the numbers injured and killed are known, nothing has been done to ameliorate the conditions. International organizations such as the ILO have issued special warnings calling attention to the problem of labor safety.[6]

Although it was widely known by the 1980s that shipbreaking had serious environmental and safety problems, there was little discussion about international measures or regulation. It was not until the 1990s, when shipbreaking moved from Taiwan to South Asia, that these problems attracted the attention of international organizations and the international environmental NGOs, and the need for action was asserted.

The framework of international regulations for the transboundary movements of hazardous wastes was enacted in the Basel Convention in 1989. The Convention may point to an expansion in the purposes for the transboundary movements for some types of recyclable materials and used equipment, other than hazardous wastes. Even if the goods dealt have value if recycled, they might have been crossing the boundary for the purpose of disposal, and the process of recycling can cause environmental pollution. The parties to the Basel Convention have therefore made the transboundary movements of wastes other than hazardous wastes the subject of debate.

For large ships intended for demolition, transfers only with the purpose of abandoning a ship in a clearly worthless state do not exist. At least part of the ship's main materials can be (and are) recycled usefully and cheaply, having been converted to a saleable form. It is not fair, and may well not even be possible, to place responsibility for a vessel's safe recycling on its original builder, because the ship may have undergone many modifications and crossed many boundaries with different owners in different trades.

Starting with this knowledge, international organizations have argued for regulation of the demolition of marine vessels and ship recycling—not only in the Conference of Parties of the Basel Convention on the Control of Transboundary Movements of Hazardous Wastes and their Disposal in UNEP, but also in the ILO and IMO. Each has published its own guidelines as a result of this debate: the Basel Convention published "Technical guidelines for the environmentally sound management of the full and partial dismantling of ships" in December 2002; the ILO published "Safety and health in shipbreaking: guidelines for Asian countries and Turkey" in October 2003; and the IMO published "The IMO guidelines on ship recycling" in December 2003. The difference between the UNEP and IMO guidelines was that the latter offer more practical advice from the ship owner's point of view.

Since none of these guidelines had legal force, however, compliance was voluntary and their effectiveness has thus been limited. In December 2005, the IMO decided to base a new treaty on the guidelines of December 2003, and from 2008 to 2009 this treaty was developed. On the basis of the perspectives mentioned above, the "Ship Recycling Convention" was adopted at an IMO diplomatic treaty conference held at Hong Kong in May 2009.

10.5   THE IMO SHIP RECYCLING CONVENTION

The IMO Ship Recycling Convention is applied to commercial ships of 500 GT, and recycling yards operating vessel demolition of those large-sized ships.[7] For large vessels an inspection and issuance of bond is required and the following provisions apply: (1) loading of toxic substances is prohibited or restricted, (2) an inventory of the toxic substances on the ship should be created and maintained, and (3) advanced preparation for the recycling of ship components must be made. The inventory is a list showing the location and estimated amounts of toxic substances in the vessel. Although inventory preparation is not required for a vessel which operates only domestically for its whole life, the inventory must be conducted if the ship is sold overseas. If it is known that a ship to be sold will be operating overseas at some time in its life, an inventory is required at the time of sale and should, therefore, have been maintained since the vessel's manufacture. However, development of inventory for naval fleet is not required, because the Ship Recycling Convention does not apply to warships and naval auxiliary.

The Convention tasks the recycling yard with providing worker safety, proper treatment and disposal of hazardous materials, and the formulation of ship recycling plans. Vessels must be disassembled and recycled only at authorized recycling yards. Flow of inventory and regulation by the Ship Recycling Convention are shown in Figure 10.2. Regulations exist for each step of the vessel's life: design, manufacture, operation, recycling preparation, and demolition. In addition to the initial and subsequent periodic inspections, the vessel also needs an international inventory bond for each of the steps listed above. For a new ship, an inventory must be created at build time. For vessels already in operation, an inventory must be created within five years after the entry-into-force of the Convention. In order for the final international bonds of inspection and recycling preparation to be issued, the inventory must be updated at the recycling preparatory step and a ship recycling plan must be drawn up in cooperation with a ship recycling yard. The recycling yard must obtain approval as a vessel recycling facility and undergo periodic inspections by the competent authorities of the country where the facility is located. The recycling yard must also prepare a ship recycling plan, in cooperation with the ship owner, at the recycling preparatory step.

In an inventory, the fundamental information about the vessel concerned should be indicated. Part I indicates the toxic substances found in the structure and equipment of the vessel. The waste produced during operation and the harmful supplies left behind in the warehouse are indicated in Parts II and III, respectively. The goods and substances to be indicated in the inventory are classified under four categories from A to D. Table A lists substances like asbestos and PCB that are prohibited or restricted. Heavy metals, radioactive materials, and other similar material appear in Table B. Table C covers lubricating oil, refrigerants, and other materials classified as potentially harmful material. Home electronics, IT apparatus, and other common consumer goods are listed in Table D.

For an inventory created during the construction of a new ship, each manufacturing supplier submits a material declaration for each step in the preparation of materials, manufacture of components, and assembly. Each component substance is carried forward from previous steps in the production chain, according to the flow of material and components, and is then compiled in the shipyard to create a definitive inventory.

The ship recycling plan drawn up at the recycling preparatory step includes an occupational safety and public health program, environmental protection program, and recycling process work plan. The occupational safety and public health program includes articles on fire prevention, safe

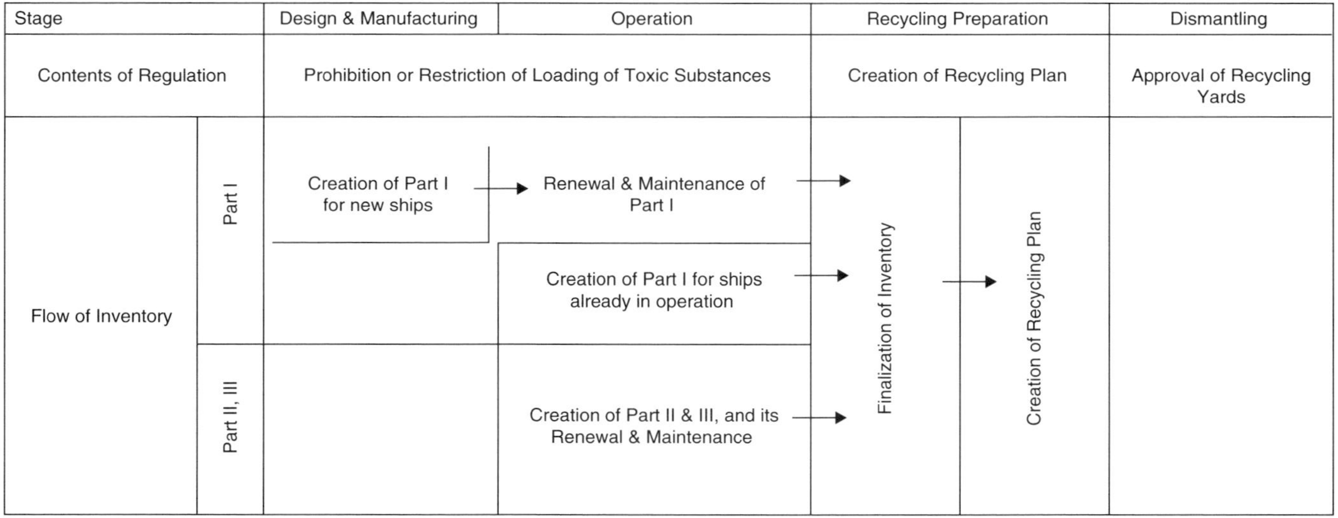

Stage	Design & Manufacturing	Operation	Recycling Preparation	Dismantling
Contents of Regulation	Prohibition or Restriction of Loading of Toxic Substances		Creation of Recycling Plan	Approval of Recycling Yards
Flow of Inventory — Part I	Creation of Part I for new ships	Renewal & Maintenance of Part I		
Flow of Inventory — Part II, III		Creation of Part I for ships already in operation		
		Creation of Part II & III, and its Renewal & Maintenance		

Finalization of Inventory → Creation of Recycling Plan

Note: The inventory is a list showing the location and estimated amounts of toxic substances in the vessel. Part I indicates toxic substances consisting in the structure and equipment of vessels. Part II indicates the waste produced during operation. Part III indicates the harmful supplies left behind in the warehouse.

Source: Based on Japan Ship Technology Research Association (2008), p. 50.

Figure 10.2   Flow of inventory and regulation by the ship recycling convention

diving operations, and the laborers' living environment; the object is to ensure worker safety and health. The measures for the management of hazardous materials, such as fuels or cargo residues and heavy metals, are included in the environmental protection program. Technical procedures for implementation of work, such as disposal of the toxic substances created during demolition, demolition itself, handling of steel scrap, worker safety, and compliance with environmental protection, are all included in the work plan.

10.6   ASSIGNMENTS OF RESPONSIBILITY IN THE DESIGN OF INSTITUTIONAL ARRANGEMENTS

Leading international environmental NGOs such as Greenpeace have repeated at every opportunity, including at the Conference of Parties of the Basel Convention, their opinion that producers and producer countries should have responsibility for the proper recycling of the vessel. Proper processing at low cost is not necessarily realized by placing the responsibility solely on the shipbuilder and the country where the ship was built, for reasons already discussed. Because of the difficulties of, for example, preparing the recycling equipment in shipbuilding countries, such policies cannot be implemented, at least in the short term.

In the IMO Ship Recycling Convention, a system in which many related subjects share responsibility at each stage was adopted. At the manufacturing stage, shipbuilders collect material declarations of raw materials, parts, and machinery from subcontractors, and create an inventory based on that information. Creation of a material declaration should be conducted by subcontractors.

After vessels are handed over, ship owners are responsible for renewing and maintaining the inventory. They also create a ship recycling plan in cooperation with recycling yards before the vessel is disassembled. Recycling yards must be approved by institutions identified by the competent authorities of the country where dismantling is carried out and undergo periodic inspections. A ship recycling plan is drawn up for every individual vessel to be demolished and suitable recycling should then be carried out.

At each stage of design, manufacture, and operation of a vessel, the competent authorities or shipping classification society of each country inspect the inventory and issue an international inventory bond. At the recycling preparatory step, just before dismantling, the final inspection of the vessel for demolition is conducted, the inventory is finalized, and a recycling preparation international bond is issued. The ship recycling plan is approved for action and a subsequent inspection is also conducted.

As mentioned above, this system for the recycling of a vessel places responsibility not only on the country of manufacture and the builder, but on every person and entity involved with building, operating, and recycling the ship. The technical characteristics of vessels are thought to have affected the creation of this recycling system, and the Scandinavian countries and Japan, historically prolific shipbuilders, are believed to have influenced the debate during negotiations for the Ship Recycling Convention in the IMO. A system placing the responsibility for recycling only on a ship's country of manufacturer has many difficulties, both technical and practical. The procedures of the Ship Recycling Convention, therefore, can be seen as being both realistic and practicable.

10.7   DISCUSSION AND CONCLUSIONS

The requirements for effectuation of the Ship Recycling Convention are that (1) 15 or more nations ratify the Convention, (2) the total merchant ship tonnage of ratifying countries will not be less than 40 percent (shipowner country provision), and (3) the lowest total maximum annual demolition in the latest ten years of ratifying destination countries does not become less than 3 percent of the merchant ship tonnage of ratifying country (demolition country provision). The Convention goes into effect 24 months after all of these requirements are met.

As 15 contracting States are needed, an agreement by the EU to adopt the Convention, which has 27 nations, by itself is sufficient. Panama and China add to the ship-owner country provision, and India will adopt as a demolition country, as will China. It is likely, therefore, that the Convention will satisfy these requirements for effectuation in the near future and that it will come into effect two years later. The Japanese government will update its own domestic law to agree with that required by the Convention before effectuation of the treaty, so that domestic law can be enforced simultaneously. In addition, at the time of effectuation of the Convention, all subject vessels will become obligated to create and maintain an inventory within five years.

It is difficult to tell whether sufficient recycling capacity for all the ships subject to the Convention can be provided. Among the demolition countries, the support of India and China were important for adoption of the Ship Recycling Convention. India, which trails Bangladesh in the demolition of large vessels, is thought to be bidding, along with China, for a leadership role by effectuating the Convention as early as possible. Bangladesh, the center of world shipbreaking at and after the beginning of the 1990s, lags behind in aligning itself with the Ship Recycling Convention. India and China recognize that the Ship Recycling Convention can provide an opportunity to expand their share of shipbreaking and promote maintenance of recycling yards to fulfill the requirements of the Convention. However, if Bangladesh, with its big share of world shipbreaking, is late in adopting the Convention, it may become difficult to secure sufficient recycling capacity once the Convention is effectuated. Major ship-owner countries, such as the EU and Japan, are concerned that the effectuation of the Ship Recycling Convention could result in a shortage in the capacity of shipbreaking yards in the world.

As of July 2013, the Ship Recycling Convention has not yet gone into effect. Moreover, none of the three conditions for effectuation have been satisfied. Among shipbreaking countries, Turkey was the first to sign in August 2010. It became the fifth signatory following France, Italy, the Netherlands, and St. Christopher and Nevis. In order to attain the demolition country provision, which is considered to be the most difficult of the three effectuation conditions, South Asian countries and China will be required to sign.

In Bangladesh in August 2010, the Supreme Court prohibited shipbreaking operations without proof of prior removal of toxic substances. However, in July 2011, the Supreme Court provisionally ruled in favor of allowing shipbreaking with the beaching system if certain conditions are fulfilled. Meanwhile, the Bangladesh government recently began designating the shipbreaking business as "an industry" to be promoted by industrial development policy. If shipbreaking is prohibited suddenly in Bangladesh, there is the possibility of a serious shortage in the demolition capacity in the world.

Although the shipbreaking business has a long history, the need to build a system for proper recycling has only recently been widely recognized. Management across borders is crucial, as it would be difficult to have individual systems for each country. The IMO Ship Recycling Convention may overcome such difficulty and is seen by many as building a realistic recycling system within a relatively short period of time.

When the history of shipbreaking on and after the mid-1960s is examined, it is clear that shipbreaking has boomed during times of high economic growth, when demand for cheap iron and steel articles increased. The business has moved from country to country as each goes through that particular stage of development. Where domestic demand for ship plate is not great, shipbreaking does not flourish. South Asia is now the center for shipbreaking activities, and can be expected to remain so while demand for ship plate for construction is high.

The complicated problem of adjusting the Basel Convention has not necessarily been fully solved, either. When the Ship Recycling Convention goes into effect, it will be necessary for the Basel Convention to enact provisions exempting marine vessels for dismantling. The Basel Convention Conference of Parties (COP) will debate the provisions, after assessment of whether the Ship Recycling Convention, as adopted, establishes an equivalent level of control and enforcement as that established under the Basel convention.

Under the Convention, the creation of the recycling system in demolition countries and the fixed capital investment by demolition contractors cannot be avoided if shipbreaking is to be carried out appropriately, with worker safety secured and environmental pollution prevented. To effectuate the Ship Recycling Convention, full cooperation of all nations—developed and developing—as well as shipbuilders, ship owners and shipbreakers, will be required.

NOTES

1. For the benefit of insurers, ship owners and other participants in maritime trade, Lloyd's Register collects data about the reduction in the total number of vessels in the world by marine accident or demolition. Gross tonnage (GT) is a unit based on the volume of a vessel and is used in the shipping industry, the main source of vessels for demolition.
2. The circumstances of the re-entry to the shipbreaking business of Japan after the 1970s are explained in detail by Sato (2004).
3. The conditions of shipbreaking in South Asian countries are based on the International Affairs Office, Ministry of Land, Infrastructure and Transport's 'Outline of Ship Recycling Convention' (2008) (in Japan Ship Technology Research Association (2008, pp. 3–10)), Sato (2004), etc.
4. Terao (2005) and Terao (2008) investigated the shipbreaking industry in Taiwan, and the factors of its development and decline.
5. Zaidanhojin Senpaku Kaitetsu Jigyo Sokushin Kyokai (Shipbreaking Business Promotion Association of Japan) (1982, pp. 102–12).
6. Extensive investigation has not been conducted on shipbreaking with the beaching system currently undertaken in India, Bangladesh, and Pakistan. Material from Greenpeace and FDIH (2005) among others exists as a report by an international environmental NGO. For ILO measures, see Sato (2004, p. 45).
7. Hereafter, explanation of the Ship Recycling Convention is mainly from the International Affairs Office, Ministry of Land, Infrastructure and Transport's (2008) "Outline of Ship Recycling Convention" (in Japan Ship Technology Research Association (2008, pp. 3–10)). For a description of the IMO (provisional) guidelines, see International Maritime Organization (2006).

REFERENCES

Greenpeace and FDIH (2005), *End of Life Ships: The Human Cost of Breaking Ships*, Amsterdam: Greenpeace International.

International Affairs Office, Ministry of Land, Infrastructure and Transport (2008), 'Outline of Ship Recycling Convention', in Japan Ship Technology Research Association (eds) (2008), (in Japanese).

International Maritime Organization (2006), *IMO Guidelines on Ship Recycling: Consolidated Edition*, London: IMO.

Japan Ship Technology Research Association (Zaidanhojin Nippon Senpaku Gijutsu Kenkyu Kyokai) (eds) (2008), *Shippu Risaikuru: Shinzosen notameno Inbentori, Zairyo Senseisho, Kyokyusha Tekigo Senseisho no Sakusei nitsuite* (Ship Recycling: On Creation of Inventory, Material Declaration, and Supplier Declaration of Conformity for Newly Built Ships), Tokyo: Japan Ship Technology Research Association (in Japanese).

Japan Ship Technology Research Association (Zaidanhojin Nippon Senpaku Gijutsu Kenkyu Kyokai) (eds) (2009), *Shippu Risaikuru: Shippu, Risaikuru Joyaku to Inbentori Sakusei nitsuite* (Ship Recycling: On the Ship Recycling Convention and Toward Creation of Inventory), Tokyo: Japan Ship Technology Research Association (in Japanese).

Lloyd's Register (various years), *Casualty Return*, London: Lloyd's Register.

Sato, Masayuki (2004), *Senpaku Kaitai: Tetsu Risaikuru Kara Mita Nippon Kindai* (Shipbreaking: History of Modern Japan from a Point of View of Iron Recycling), Tokyo: Kadensha (in Japanese).

Shipbuilder's Association of Japan (eds) (2009), *Shipbuilding Statistics 2009*, Tokyo: Shipbuilder's Association of Japan.

Terao, Tadayoshi (2005), 'The rise and fall of "mixed metal scrap" recovery industry in Taiwan: international trade of scraps and transboundary relocation of the business', in Michikazu Kojima (ed.), *International Trade of Recyclable Resources in Asia* (IDE Spot Survey No. 29), Chiba: Institute of Developing Economies-JETRO, pp. 63–84.

Terao, Tadayoshi (2008), 'Shipbreaking and metal recycling industries in Taiwan', in Michikazu Kojima (ed.), *Promoting 3Rs in Developing Countries: Lessons from Japanese Experiences* (IDE Spot Survey No. 30), Chiba: Institute of Developing Economies-JETRO, pp. 59–79.

Zaidanhojin Senpaku Kaitetsu Jigyo Sokushin Kyokai (Shipbreaking Business Promotion Association of Japan) (1982), *Kaitetsu Senka no Chosa Bunseki* (Research and Analysis on Prices of Vessels for Demolition), Tokyo: Zaidanhojin Senpaku Kaitetsu Jigyo Sokushin Kyokai (in Japanese).

11. Toward efficient resource utilization in the Asian region

Michikazu Kojima

One of the characteristics of economic integration in Asia is the fragmentation of various production processes. Not only the final products, but also raw materials and intermediate goods are traded among the Asian countries. Recyclable waste generated from producers and consumers is also traded internationally. Asia has imported various types of recyclable wastes from other regions because currently it is the production center of the world, utilizing recyclable waste as resources (see Chapter 2).

It has been pointed out that various factors such as low labor cost, externalization of environmental protection cost, and trade regulation on recyclable waste have affected the flow of recyclable waste and location of the recycling industry (see Chapters 3 and 4 of this book, and Kojima (ed.) 2005). In addition, the import duty reduction system has impacted trade flows of recyclable waste (Chapter 8), and the import protection of steel products has led to the import of waste ships (Chapter 10).

In the Asian region, there are many informal recyclers that do not have environmentally sound technology. They often have price competitiveness because of their lower pollution control cost compared with formal recyclers with environmentally sound technology. As a result, waste generators send their waste to informal recyclers at a higher price or with lower treatment cost. Such circumstances discourage investment in environmentally sound recycling. In Asia, informal recyclers and formal recyclers are competing throughout the region: formal recyclers in a country are competing with informal recyclers in other countries. As a result, environmental problems resulting from the recycling of imported waste have become social issues in importing countries, while the shortage of recyclable materials has become an issue in exporting countries.

Some countries, such as China (Chapter 3), Vietnam (Chapter 4), and South Korea (Chapter 5), which have suffered from improper import and recycling of recyclable waste and hazardous waste, are struggling to minimize the negative impacts from importation and processing of the recyclable waste and hazardous waste. Countermeasures taken by these countries include bans on imports, permits for importers, permits for exporters, quality standards for imported recyclable waste, prior notice and consent schemes for hazardous waste, among other strategies. These measures are effective in preventing the transboundary movement of waste destined for improper disposal and recycling. However, unilateral measures often create conflict with other countries (see Chapters 4 and 9). If the measures are implemented with the cooperation of importing and exporting countries, they become more effective.

Asian countries also have realized gradually that they are becoming exporters of recyclable waste and hazardous waste. They are facing new challenges, especially on how to establish societies with a sound material cycle or recycling system in connection with the export of recyclable waste and hazardous waste. For example, the Philippines has enacted a comprehensive act to enhance integrated solid waste management and recycling. Many material recovery centers have been established to divert waste from landfills to recycling. However, a lot of collected recyclable waste has been exported to other countries. The country faces new challenges to take into account international factors in order to close the loop of material recycling (Chapter 6). Some countries such as South Korea (Chapter 5) and Japan (Chapter 7) are struggling to control improper export of hazardous waste. Such unilateral actions in an exporting country also need cooperation from the importing country. For example, information on improper recycling in an importing country can be used for prioritizing problematic waste, for justifying policy measures in the exporting country, and for identifying the flow of waste.

The establishment of a society with a sound material cycle and recycling based system cannot be completed in a single country in Asia, because of the fragmentation of the recycling process, along with economic integration in the region. However, most countermeasures to control illegal inflow and outflow of hazardous and other waste are unilateral actions. The only mechanism available, which covers most of the Asian countries, is the Basel Convention. It defines the obligation and the right of countries to control the transboundary movement of hazardous waste (Chapters 1 and 9), but each country is allowed to create its own definition of hazardous waste. The differences in definition of hazardous waste and a lack of information dissemination on the definitions often cause unintentional illegal transboundary movement and conflict between parties (Chapter 9).

It is time for Asian countries to determine and introduce more comprehensive cooperation mechanisms to promote efficient use of resources, by preventing illegal transboundary movement of hazardous waste and by promoting the international trade of recyclable waste and hazardous waste destined for recycling with environmentally sound technology.

Harmonizing definitions of hazardous waste can be a basis for such regional effort. However, there are many differences in definitions and related standards, such as testing methods and thresholds of hazardous substances. The first step might be information sharing on the definitions of hazardous waste and strengthening the joint enforcement on illegal transboundary movement of hazardous waste. Joint inspection by several agencies among the countries in Asia could lead to more trust between the countries and to more effective control measures. On the other hand, companies with high environmental awareness and good environmental technology are negatively affected by high transaction costs, including documentation needs for prior notice and consent. Government should simplify the prior notice and consent procedures for recyclers with environmentally sound technology to enhance the proper international trade of hazardous waste and discourage improper international trade indirectly.

Ship recycling needs a different international scheme to ensure environmentally sound management. Used ships and ships destined for recycling are traded differently from other goods. Often the country of the ship owner and the registration country of the ship are different. Controlling the transboundary movement of waste ships by using the control scheme of the Basel Convention remains difficult. The government to which a ship has been registered cannot control changes in ownership or registration country. For these reasons, the Ship Recycling Convention is being established. Over the past several decades, the major work in dismantling ships has migrated among Asian countries, such as from Japan to Taiwan in the 1970s, and from Taiwan to Bangladesh, India, Pakistan, and China in the 1990s. The reasons behind this migration include not only cheap labor costs and lack of environmental protection measures, but also high demand for second-hand goods and higher tariffs on steel products. In addition to the Ship Recycling Convention, elimination of higher tariffs on steel products may discourage dirty recycling of ships (Chapter 10).

One of the key features of the Ship Recycling Convention is the certification system of environmentally sound ship dismantling facility and process. A similar system can be applied to transboundary movement of hazardous waste, under the jurisdiction of the Basel Convention. In the COP 10 of the Basel Convention, parties decided to work on environmentally sound management. The work on environmentally sound management (ESM) by the Convention is expected to result in facilitating investment in ESM facilities in developing countries and enhancing the proper international trade of recyclable waste and hazardous waste.

When we discussed the issue of transboundary movement of hazardous and recyclable waste with staff of NGOs and government officials in importing countries, we often encountered their distrust of exporting countries. They claim that exporting countries dump their waste on the importing country, which causes environmental problems. On the other hand, when we discussed the issue with recycling companies and government officials in exporting countries, we often encountered distrust of importing countries. They claim that export of recyclable waste and smuggling of hazardous waste undermines their effort to establish a sound material cycle. However, Asia as a whole in fact faces the challenge of how to establish a society with a sound material cycle and recycling system in the region. As mentioned in Chapter 1, there are few studies on the transboundary movement of hazardous waste and recyclable waste. We hope this book facilitates mutual understanding among Asian countries and serves as a basis for international and regional cooperation to establish society with a sound material cycle, not just in Asia but worldwide.

REFERENCE

Kojima, Michikazu (ed.) (2005), *International Trade of Recyclable Resources in Asia*, Institute of Developing Economies, http://www.ide.go.jp/English/Publish/Spot/29.html.

Index